T0138115

Communications
in Computer and Information Science   2021

Rationale

The CCIS series is devoted to the publication of proceedings of computer science conferences. Its aim is to efficiently disseminate original research results in informatics in printed and electronic form. While the focus is on publication of peer-reviewed full papers presenting mature work, inclusion of reviewed short papers reporting on work in progress is welcome, too. Besides globally relevant meetings with internationally representative program committees guaranteeing a strict peer-reviewing and paper selection process, conferences run by societies or of high regional or national relevance are also considered for publication.

Topics

The topical scope of CCIS spans the entire spectrum of informatics ranging from foundational topics in the theory of computing to information and communications science and technology and a broad variety of interdisciplinary application fields.

Information for Volume Editors and Authors

Publication in CCIS is free of charge. No royalties are paid, however, we offer registered conference participants temporary free access to the online version of the conference proceedings on SpringerLink (http://link.springer.com) by means of an http referrer from the conference website and/or a number of complimentary printed copies, as specified in the official acceptance email of the event.

CCIS proceedings can be published in time for distribution at conferences or as postproceedings, and delivered in the form of printed books and/or electronically as USBs and/or e-content licenses for accessing proceedings at SpringerLink. Furthermore, CCIS proceedings are included in the CCIS electronic book series hosted in the SpringerLink digital library at http://link.springer.com/bookseries/7899. Conferences publishing in CCIS are allowed to use Online Conference Service (OCS) for managing the whole proceedings lifecycle (from submission and reviewing to preparing for publication) free of charge.

Publication process

The language of publication is exclusively English. Authors publishing in CCIS have to sign the Springer CCIS copyright transfer form, however, they are free to use their material published in CCIS for substantially changed, more elaborate subsequent publications elsewhere. For the preparation of the camera-ready papers/files, authors have to strictly adhere to the Springer CCIS Authors' Instructions and are strongly encouraged to use the CCIS LaTeX style files or templates.

Abstracting/Indexing

CCIS is abstracted/indexed in DBLP, Google Scholar, EI-Compendex, Mathematical Reviews, SCImago, Scopus. CCIS volumes are also submitted for the inclusion in ISI Proceedings.

How to start

To start the evaluation of your proposal for inclusion in the CCIS series, please send an e-mail to ccis@springer.com.

Mamata Dalui · Sukanta Das · Enrico Formenti

Editors

Cellular Automata Technology

Third Asian Symposium, ASCAT 2024
Durgapur, India, February 29 – March 2, 2024
Revised Selected Papers

Springer

Editors
Mamata Dalui (iD)
National Institute of Technology Durgapur
Durgapur, India

Enrico Formenti (iD)
Université Côte d'Azur
Nice, France

Sukanta Das (iD)
Indian Institute of Engineering Science
and Technology
Shibpur Howrah, West Bengal, India

ISSN 1865-0929   ISSN 1865-0937 (electronic)
Communications in Computer and Information Science
ISBN 978-3-031-56942-5   ISBN 978-3-031-56943-2 (eBook)
https://doi.org/10.1007/978-3-031-56943-2

This Springer imprint is published by the registered company Springer Nature Switzerland AG
The registered company address is: Gewerbestrasse 11, 6330 Cham, Switzerland

Paper in this product is recyclable.

Preface

A warm welcome to the proceedings of Third Asian Symposium on Cellular Automata Technology, 2024 (ASCAT 2024), a gathering that brought together researchers, academicians, and professionals at the forefront of exploring the intricate world of cellular automata. This symposium serves as a platform for the exchange of ideas, insights, and innovations in the fascinating realm of computational and mathematical models. It investigates cellular automata as a technology and fosters cellular automata-related theories. Therefore, the symposium encompasses all theoretical facets of cellular automata and their practical implementations across various domains. The applications of cellular automata span a multitude of fields, from computer engineering to transportation engineering to material science, and from physics to biology and social science. The symposium aims to explore the latest advancements, methodologies, and interdisciplinary connections that showcase the versatility and potential impact of cellular automata in contemporary research.

ASCAT 2024 was the third edition of the symposium series, which actually started in the year 2022. Both the first (ASCAT 2022) and second (ASCAT 2023) editions were hosted by Indian Institute of Engineering Science and Technology, Shibpur (IIEST, Shibpur), India. This time the event was officially hosted by the Department of Computer Science and Engineering, National Institute of Technology, Durgapur (NIT, Durgapur) at its Campus at West Bengal, India in hybrid mode during February 29 – March 2, 2024.

This year we had the pleasure to have amongst us eminent researchers like Kenichi Morita (Hiroshima University, Japan), R. Ramanujam (IMSc, Chennai, India), Niloy Ganguly (IIT, Kharagpur, India), Genaro Juarez Martinez (National Polytechnic Institute, Mexico), Nazim Fatès (Inria-Loria, France), Rezki Chemlal (University of Bejaia, Algeria), Manoj Nambiar (TCS Research, India) and Rajesh P. Barnwal (CMERI, India) as the Invited Speakers of ASCAT 2024. Our distinguished invited speakers and presenters shared their cutting-edge research, shedding light on new perspectives, methodologies, and applications in the relevant field. On behalf of the organising committee of ASCAT 2024, we express our sincere gratitude to all the speakers.

The current volume of proceedings includes one invited paper (the first paper of these proceedings) and 15 contributed papers accepted for presentation at ASCAT 2024. This year we received a total of 33 high-quality papers, which were rigorously reviewed, primarily by the Program Committee (PC) members. Each submission received at least 3 reviews were single blind. After the review phase, all the reviews were made open to the PC members for discussion. Finally, in an online PC meeting, 9 papers were accepted for presentation and 7 papers were accepted provisionally subject to compliance with the reviewers' comments. The provisionally accepted papers were revised and reviewed again, and finally 6 of them were accepted for presentation at ASCAT 2024. We are indebted to our reviewers for their timely effort and commitment in reviewing the submissions rigorously. We extend our appreciation to every author for their contributions and diligent effort that made it possible to successfully execute the event. We are also grateful to Kenichi Morita, who responded to our invitation and submitted an invited article.

We express our sincere gratitude to the members of the programme committee for efficiently executing this substantial volume of tasks within the constrained time frame. The process of paper submission, the entire review process and notifications were managed through the EasyChair conference management system. We are thankful to the service provider for it. We would like to extend our heartfelt gratitude to the Springer representatives who consented to publish the proceedings of ASCAT 2024 in the Springer CCIS Series.

A research group, "Cellular Automata India", (https://www.cellularautomata.in/) which was founded during the COVID 19 pandemic, played a pivotal role in ensuring the success of the event. As a regular activity of this group, a summer school on cellular automata, named Summer School on Cellular Automata Technology 2023, was organized during the summer vacation (May-July) of 2023. This was the third edition of the summer school, which was co-organised by University of Bejaia, Algeria, where undergraduate/postgraduate students, Ph.D. scholars/researchers and faculty members actively participated. Nine submissions were received for ASCAT 2024 as an outcome of the research projects that were initiated during the summer school, out of which six were accepted for presentation.

The success of the symposium would not have been possible without the generous contributions, resources, and collaborative spirit of the Department of Computer Science and Engineering, National Institute of Technology Durgapur. We express our appreciation to the entire team of volunteers, sponsors, and participants who contributed to the success of this event. Last but not least, we are ever grateful to National Institute of Technology, Durgapur for entrusting us with the opportunity to host ASCAT 2024 at its campus.

February 2024

Suchismita Roy
Biplab K. Sikdar
Kenichi Morita
General Co-chairss

Mamata Dalui
Sukanta Das
Enrico Formenti
Program Co-chairss

Organization

Chief Patron

Arvind Choubey (Director) National Institute of Technology Durgapur, India

General Co-chairs

Suchismita Roy National Institute of Technology Durgapur, India
Biplab K. Sikdar Indian Institute of Engineering Science and
  Technology, Shibpur, India
Kenichi Morita Hiroshima University, Japan

Programme Co-chairs

Mamata Dalui National Institute of Technology Durgapur, India
Sukanta Das Indian Institute of Engineering Science and
  Technology, Shibpur, India
Enrico Formenti Université Côte d'Azur, France

Program Committee

Sumit Adak Technical University of Denmark, Denmark
Kamalika Bhattacharjee NIT Tiruchirappalli, India
Pabitra Pal Chaudhuri ISI, Kolkata, India
Rezki Chemlal University of Bejaia, Algeria
Mamata Dalui NIT Durgapur, India
Sukanta Das IIEST Shibpur, India
Enrico Formenti Université Côte d'Azur, France
Raju Hazari NIT Calicut, India
Huynh Xuan Hiep Can Tho University, Vietnam
Teijiro Isokawa University of Hyogo, Japan
Jimmy Jose NIT Calicut, India
Supreeti Kamilya BIT Mesra, India
Anna Lawniczak Guelph University, Canada
Genaro Juarez Martinez National Polytechnic Institute, Mexico

Kenichi Morita	Hiroshima University, Japan
Sukanya Mukherjee	IEM, Kolkata, India
Nazma Naskar	KIIT University, India
Stefano Nichele	Oslo Metropolitan University, Norway
Pedro Paulo Balbi de Oliveira	Universidade Presbiteriana Mackenzie, Brazil
Souvik Roy	Ahmedabad University, Gujarat, India
Suchismita Roy	NIT Durgapur, India
Sudhakar Sahoo	IMA, Bhubaneswar, India
Bibhash Sen	NIT Durgapur, India
Biplab K. Sikdar	IIEST Shibpur, India
Hiroshi Umeo	Osaka Electro-Communication University, Japan

Additional Reviewers

Bidesh Chakraborty
Pedro Costa
Baisakhi Das
Nilanjana Das
Suvadip Hazra
Luca Manzoni
Surajit Kumar Roy
Mousumi Saha
Sutapa Sarkar

Organizing Co-chairs

Bibhash Sen	National Institute of Technology Durgapur, India
Ankush Archarya	National Institute of Technology Durgapur, India
Suvadip Batabyal	National Institute of Technology Durgapur, India

Publication Co-chairs

Sujan Saha	National Institute of Technology Durgapur, India
Souvik Roy	Ahmedabad University, Gujarat, India
Mousumi Saha	National Institute of Technology Durgapur, India

Publicity Co-chairs

Kamalika Bhattacharyya	National Institute of Technology Tiruchirappalli, India
Shayantani Maity	National Institute of Technology Durgapur, India
Debasis Mitra	National Institute of Technology Durgapur, India

Registration Co-chairs

Parag K. Guhathakurta	National Institute of Technology Durgapur, India
Subhas Sahana	National Institute of Technology Durgapur, India
Monalisa Mondal	National Institute of Technology Durgapur, India

Local Organizing Committee

Biplab K. Sikdar	IIEST Shibpur, India
Suchismita Roy	NIT Durgapur, India
Sukanta Das	IIEST Shibpur, India
Bibhash Sen	NIT Durgapur, India
Parag K. Guhathakurta	NIT Durgapur, India
Ankush Acharyya	NIT Durgapur, India
Suvadip Batabyal	NIT Durgapur, India
Sujan Saha	NIT Durgapur, India
Sukanya Mukharjee	IEM, Kolkata, India
Kamalika Bhattacharjee	NIT Tiruchirappalli, India
Mamata Dalui	NIT Durgapur, India

Contents

Making and Using a Rotary Element in Reversible Cellular Automata

Kenichi Morita(✉)⑩

Hiroshima University, Higashi-Hiroshima 739-8527, Japan
km@hiroshima-u.ac.jp

Abstract. A rotary element (RE) is a reversible logic element with one-bit memory (RLEM) proposed by Morita (2001). Though it is a very simple RLEM and its operation is easily understood, it is universal in the sense that any other RLEM is composed only of it. In this survey, we discuss how we can compose an RE in simple reversible cellular automata (RCAs), and how it is used to show universality of RCAs. Here, we consider two examples of RCAs, which are reversible elementary square partitioned CAs (ESPCAs). We first explain that in each of these RCAs, an RE is implemented using only a few kinds of small patterns and their interactions. We then discuss how an RE is used to show Turing universality and intrinsic universality of RCAs. Turing universality of RCAs is derived by constructing reversible Turing machines (RTMs) out of the implemented RE. Utilizing an RE, we can also show intrinsic universality of the RCAs, which is the property of a particular RCA that any RCA in some large class of RCAs can be simulated in it. Computing processes of them can be seen on the CA simulator Golly.

Keywords: rotary element · reversible logic element with memory · reversible cellular automaton · Turing universality · intrinsic universality · Golly

1 Introduction

Reversible computing is a paradigm that reflects physical reversibility. Since reversibility is one of the fundamental microscopic laws of nature, it is important to investigate how higher functions, such as a universal computing capability, emerge from a simple reversible microscopic law. Here, we study this problem using reversible cellular automata (RCAs) that obey simple reversible local transition rules.

When we compose a universal computer in a reversible CA, it is convenient to put a conceptual device between the microscopic level corresponding to simple local rules, and the macroscopic level corresponding to a universal computer. A reversible logic gate, such as a Fredkin gate [2] and a Toffoli gate [14], is one of the candidates. However, here, we use a reversible logic element with memory

---

K. Morita—Currently Professor Emeritus of Hiroshima University.

M. Dalui et al. (Eds.): ASCAT 2024, CCIS 2021, pp. 1–16, 2024.
https://doi.org/10.1007/978-3-031-56943-2_1

(RLEM) as a basic device, since it is useful for composing larger reversible systems very simply [3,5,8]. Among RLEMs, a *rotary element* (RE) [3] is typical one with two states, whose operations are easily understood. Furthermore, it is universal in the sense that any RLEM can be realized by a circuit composed only of REs.

In this survey, we show how we can implement an RE in simple reversible CAs, and how it is utilized to compose complex reversible systems. Here, we use the framework of an elementary square partitioned CA (ESPCA). A cell of an ESPCA has four parts each of which has two states, and thus has 16 states in total. Its local function is defined by only six local transition rules, and hence very simple. We consider two kinds of reversible ESPCAs with identification numbers 01caef and 02c5bf as examples. First, we show composing methods of an RE in these reversible ESPCAs. We can see that, in each of these ESPCAs, an RE is implemented using only two kinds of small patterns and a few kinds of their interaction processes. We then show Turing universality and intrinsic universality of these reversible ESPCAs. Utilizing the implemented RE, we can explicitly give a configuration that simulates a given reversible Turing machine (RTM), and a configuration that simulates another reversible ESPCA. Full evolution processes of these configurations can be seen by running emulators [7,10] that are executable on the general purpose CA simulator *Golly* [15].

The organization of the paper is as follows. In Sect. 2, a rotary element (RE) is introduced, and its intrinsic universality is explained. In Sect. 3, the framework of elementary square partitioned cellular automata (ETPCAs) is given. Then, construction methods of an RE in two kinds of particular reversible ESPCAs are described. In Sect. 4, Turing universality and intrinsic universality of the two ESPCAs are shown by utilizing the composed REs. In Sect. 5, concluding remarks are given.

Note that, in this survey, only outlines of the construction methods are explained. For their details, see the references given in each section. References [5,8,11] describe RLEMs, ESPCAs and RTMs, generally.

2   Rotary Element (RE)

A *rotary element* (RE) is a 2-state device having four input ports and four output ports [3]. Conceptually, it has a rotatable bar inside as shown in Fig. 1. Its two states are distinguished by the direction of the bar. They are the state V and the state H as depicted in the figure.

Its operation is as follows. If a signal comes from the direction parallel to the bar, then it goes straight ahead, and the state does not change (Fig. 2, left). If a signal comes from the direction orthogonal to the bar, then it makes a right turn, and the state changes to the other (Fig. 2, right). An RE is reversible in the following sense: From the next state and the output, the previous state and the input are uniquely determined. Note that it is not allowed to give two or more input signals at the same time to an RE.

We now give a definition of a reversible sequential machine (RSM), since a reversible logic element with memory (RLEM) is a kind of an RSM.

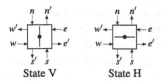

Fig. 1. Two states of a rotary element (RE) [3]

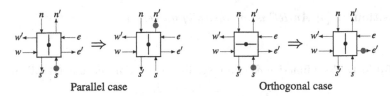

Parallel case   Orthogonal case

Fig. 2. Operations of an RE [3]

Definition 1. *A sequential machine (SM) M is defined by $M = (Q, \Sigma, \Gamma, \delta)$, where Q is a finite set of states, Σ and Γ are finite sets of input and output symbols, and $\delta : Q \times \Sigma \to Q \times \Gamma$ is a move function. If δ is injective, it is called a* reversible sequential machine *(RSM).*

A *reversible logic element with memory* (RLEM) is an RSM that satisfies $|\Sigma| = |\Gamma|$. If $|Q| = n$ and $|\Sigma| = |\Gamma| = k$, it is called an *n-state k-symbol RLEM*. Thus, an RE is a 2-state 4-symbol RLEM.

Definition 2. *An RLEM M is called* intrinsically universal, *if any RSM (hence any RLEM) is realized by a circuit composed only of M.*

An RE has been shown to be intrinsically universal [4,5]. We explain it by an example.

Let M_0 be a 3-state 3-symbol RSM whose move function δ_0 is given in Table 1. It is realized by an RE circuit shown in Fig. 3. It consists of 12 REs, which are arranged in a 4×3 array. We denote the RE on the i-th row of the j-th column by (i, j)-RE. The three columns of REs correspond to the states q_1, q_2 and q_3 of M_0. The i-th row ($i \in \{1, 2, 3\}$) of REs corresponds to the input c_i and the output d_i. If M_0 is in q_j ($j \in \{1, 2, 3\}$), then the $(4, j)$-RE is set to the state H. All the other REs are set to the state V. Assume $\delta_0(q_j, c_i) = (q_{j'}, d_{i'})$. Also assume the state of the circuit is now q_j, and an input is given to the port c_i. Then, we can see that the circuit finally becomes the state $q_{j'}$ and the signal comes out from $d_{j'}$.

For example, assume M_0 is in q_0 and an input signal is given to the port c_1 as in Fig. 3. The signal first changes the state of the $(1, 1)$-RE to H, and goes down in the 1st column. Next, the signal changes the state of the $(4, 1)$-RE to V, and then restores the state of the $(1, 1)$-RE to V. The signal travels from the west output port of the $(1, 1)$-RE to the east input port of the $(2, 2)$-RE. It changes the states of the $(2, 2)$-RE and the $(4, 2)$-RE to H. It then restores the state of the $(2,2)$-RE into V, and comes out from its east output port. In the third column, the signal temporarily changes the state of the $(2,3)$-RE to H, and goes down. It makes a U-turn at the $(4,3)$-RE, and restores the state of the $(2,3)$-RE. Finally, the signal goes out from the output port d_2. By the above process, the transition $\delta_0(q_1, c_1) = (q_2, d_2)$ is simulated.

In a similar way, any RSM can be realized by an RE circuit. Thus, we have the following proposition.

Proposition 1. [4] *An RE is intrinsically universal.*

Table 1. Move function δ_0 of a reversible sequential machine (RSM) M_0

	Present state		
Input	q_1	q_2	q_3
c_1	(q_2, d_2)	(q_3, d_2)	(q_3, d_3)
c_2	(q_3, d_1)	(q_2, d_1)	(q_2, d_3)
c_3	(q_1, d_2)	(q_1, d_3)	(q_1, d_1)

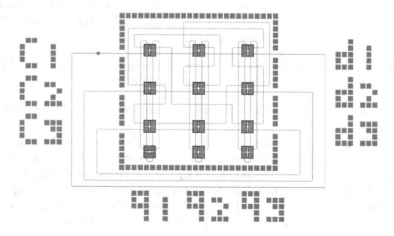

Fig. 3. RSM M_0 realized by a circuit of REs. It is simulated on Golly [10]

3   Making an RE in Simple RCAs

We shall see that an RE is realized in very simple reversible CAs. Here, we use elementary square partitioned cellular automata to do so. They are the CAs first introduced in [12].

3.1   Elementary Square Partitioned Cellular Automaton (ESPCA)

An *elementary square partitioned cellular automaton* (ESPCA) is a 2-dimensional CA whose cell is divided into four parts (Fig. 4(a)). Each part has the state set $\{0, 1\}$, and thus a cell has 16 states in total. A cell changes its state by a *local function*. It is defined by a set of *local transition rules* of the form shown in Fig. 4(b). In an ESPCA, we assume that the local function is rotation-symmetric. Therefore, it is described by six local transition rules as in Fig. 6. We use the method of representing a local function by a 6-digit hexadecimal number as given in Fig. 5. Thus, the local function of Fig. 6 has the identification number 01caef. Applying a local function to all the cells in parallel, we have a *global function*. By it, configurations (*i.e.*, whole states of the cellular space) evolve. See *e.g.*, [5,8,11] for the details of the definition.

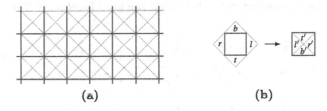

(a)   **(b)**

Fig. 4. (a) Cellular space of an elementary square partitioned cellular automaton (ESPCA), and (b) local transition rule of an ESPCA

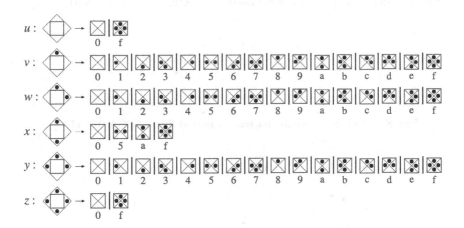

Fig. 5. Representing an ESPCA by a 6-digit hexadecimal number $uvwxyz$

An ESPCA is called *reversible* if its global function is injective. In an ESPCA, it is reversible if and only if its local function is injective (see e.g., [5]). Hence, we can obtain a reversible ES'PCA easily.

3.2 Making an RE in ESPCA-01caef

We first consider ESPCA-01caef. Its local function is in Fig. 6. It is injective, since there is no pair of distinct local transition rules that have the same right-hand sides. In [6], it was used to compose another RLEM with the identification number 4-31. Here, we show that an RE is also implemented in the same ESPCA, though the resulting configuration is more complex than the case of RLEM 4-31.

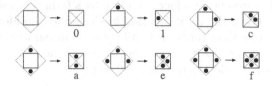

Fig. 6. Six local transition rules that define the local function of ESPCA-01caef

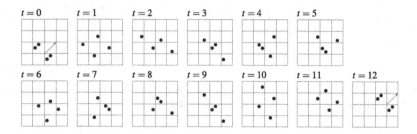

Fig. 7. Glider-12, a space-moving pattern of period 12 in ESPCA-01caef [6]

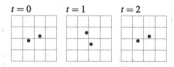

Fig. 8. Blinker, a periodic pattern of period 2 in ESPCA-01caef [6]

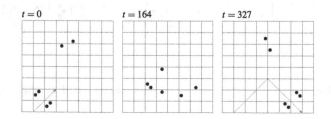

Fig. 9. Right-turn of a glider-12 in ESPCA-01caef [6]

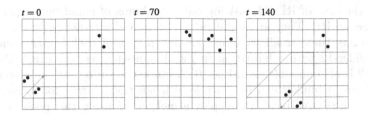

Fig. 10. U-turn of a glider-12 in ESPCA-01caef [6]

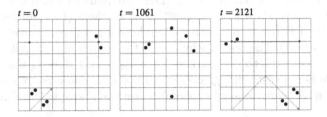

Fig. 11. Shifting a blinker by a glider-12 in ESPCA-01caef [6]

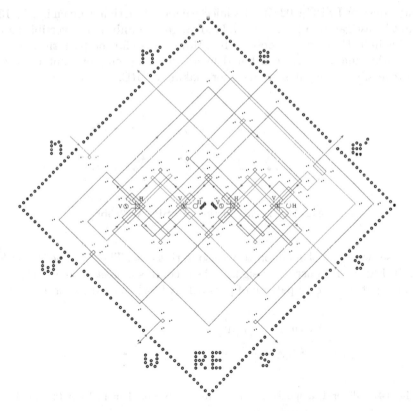

Fig. 12. RE implemented in ESPCA-01caef [11]

As in the case of RLEM 4-31 [6], only two kinds of small patterns are used to compose an RE. They are a *glider-12* (Fig. 7) and a *blinker* (Fig. 8).

Interacting these small patterns, we obtain several useful phenomena. Colliding a glider-12 with a blinker as in Fig. 9 right-turn is performed. U-turn of a glider-12 is also possible as in Fig. 10. A blinker is shifted by a glider-12 as in Fig. 11. Thus, a blinker can play a role of a *position marker* for memorizing a position. In ESPCA-01caef, we employ a method of using this position marker for keeping a state of an RE.

By the three phenomena in Figs. 9, 10 and 11, we can compose a pattern that simulates an RE. It is given in Fig. 12. The four pairs of small circles near the center of the pattern show possible position of markers (*i.e.*, blinkers). If every marker is in the left (right, respectively) circle of a pair, we assume the RE is in the state V (state H). The reason why we use four markers is to make a sufficient number of access paths to the markers for sensing and shifting the markers by a glider-12. When a state change occurs, all the four markers are shifted to the same direction. A more detailed explanation on the above is found in [11].

3.3   Making an RE in ESPCA-02c5bf

Second, consider ESPCA-02c5bf. Its injective local function is given in Fig. 13. It was first investigated in [12], where a Fredkin gate, a universal reversible gate, is realizable in it. Here, we compose an RE. However, so far, no pattern that works as a position marker has been found in it. Thus, we employ another method where reversible logic gates are used for making an RE.

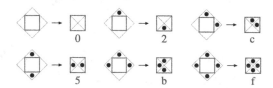

Fig. 13.   Local function of ESPCA-02c5bf

We use two kinds of small patterns also in this case. The first one is a *glider-1* shown in Fig. 14. It is used as a signal. The second is a *block* shown in Fig. 15. It is a stable pattern, *i.e.*, a pattern of period 1. A block is used to control a signal.

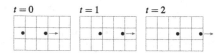

Fig. 14.   Glider-1, a space-moving pattern of period 1 in ESPCA-02c5bf [12]

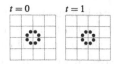

Fig. 15. Block, a stable pattern in ESPCA-02c5bf [12]

We use two kinds of phenomena for composing an RE. The first one is shown in Fig. 16. Colliding a glider-1 with a block, left-turn of a signal is realized. Note that right-turn is performed by three left-turns. Second, colliding two signals as in Fig. 17, their trajectories shift by one cell. By this, we obtain a kind of reversible logic gate called an *interaction gate* (I-gate) [2] (Fig. 18(a)). To make an RE, we also need an *inverse interaction gate* (I^{-1}-gate) (Fig. 18(b)). Since I^{-1}-gate is a 4-input 2-output gate, it is a partially defined reversible gate. In ESPCA-02c5bf, it is obtained by reversing the moving directions of the signals in Fig. 17.

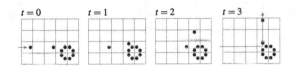

Fig. 16. Left-turn of a signal in ESPCA-02c5bf [12]

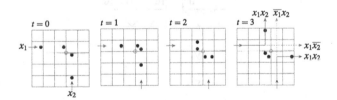

Fig. 17. Interaction of two signals in ESPCA-02c5bf [12].

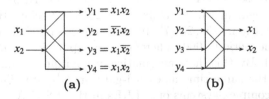

Fig. 18. (a) Interaction gate (I-gate), and (b) its inverse gate (I^{-1}-gate) [2]

We can compose an RE out of I-gates, I^{-1}-gates, and delay elements as shown in Fig. 19 [9]. Small triangles in this figure are delay elements, in which delay time is written. Since a logic gate is a memory-less device, we have to provide a circulating signal called a *state signal* to keep the state of an RE. The cycles labelled V and H are the ones for this purpose. If the state signal circulates along the cycle labelled V (H, respectively), we assume the RE is in the state V (H). In Fig. 19, the period of a state signal is 4, and thus an input signal must be given at $t \equiv 0 \pmod 4$. We can see that this circuit correctly simulates an RE. A detailed explanation is found in [9].

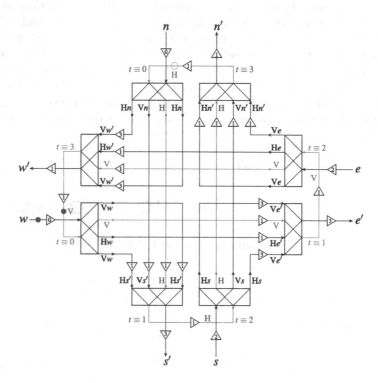

Fig. 19. RE implemented by I-gates, I^{-1}-gates, and delay elements [9]

Using the processes shown in Figs. 16 and 17 properly, an I-gate, an I^{-1}-gate and a delay element can be realized in ESPCA-02c5bf. By them, an RE is obtained as shown in Fig. 20 [7]. In this implementation, the state signal circulates with the period 1000. Therefore, an input signal must be given at $t \equiv 0 \pmod{1000}$. By this, it is automatically guaranteed that signals always arrive at exactly the same time at each I-gate or I^{-1}-gate. This fact makes it easy to compose complex systems out of REs in this ESPCA.

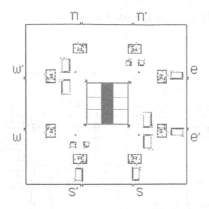

Fig. 20. RE implemented in ESPCA-02c5bf simulated on Golly [7]

4 Utilizing an RE for Showing Universality of RCAs

We use an RE to show two kinds of universality of reversible ESPCAs. *Turing universality* of an RCA is the property where any Turing machine (TM) can be simulated in its cellular space. See, *e.g.*, [5,11] for the details of the definition. Note that it has been shown that any TM can be simulated by a garbage-less reversible TM (RTM) by Bennett [1]. Therefore, to prove Turing universality of an RCA, it is sufficient to show that it can simulate any RTM.

If a universal logic element such as an RE or a Fredkin gate is implemented in a reversible ESPCA, we can conclude its Turing universality. This is because a finite control and a tape cell of an RTM are formalized as a reversible sequential machine (see *e.g.*, [5,8,11]), and thus we can, in principle, construct any RTM in the ESPCA. However, if we use only reversible gates, its construction will become very complex, since there is a signal timing problem at each gate. On the other hand, if an RE is implemented in ESPCAs, it becomes much easier to explicitly construct RTMs in these ESPCAs. Furthermore, by the CA simulator Golly [15], we can see full computing processes of them [7,10].

There is another kind of universality of a CA called *intrinsic universality*. It is the property of a CA where any CA in a large class of CAs can be simulated in the former CA. See, *e.g.*, [5,11,13] for the details of the definition. Below, we shall see that the RE patterns in the reversible ESPCAs are also useful for explicitly constructing configurations that simulate other ESPCAs.

4.1 Turing Universality of RCAs

It has been shown that any RTM can be composed of REs easily in a systematic manner [3]. A simple example of an RTM T_{parity} constructed out of REs is shown in Fig. 21. The composing method was first given in [3], and revised in [8]. The RTM T_{parity} accepts the unary language $\{1^n \mid n$ is even$\}$ and has the following set of quintuples: $\{[q_0, 0, 1, R, q_1], [q_1, 0, 1, L, q_a], [q_1, 1, 0, R, q_2],$

$[q_2, 0, 1, L, q_r], [q_2, 1, 0, R, q_1]\}$. A detailed explanation of the circuit of Fig. 21 is found in [8,11].

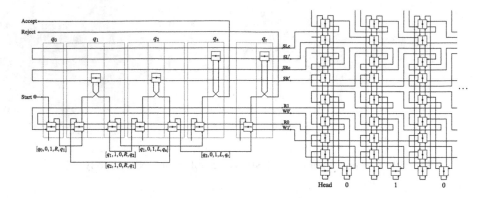

Fig. 21. RTM T_{parity} composed of REs [8]

Replacing each occurrence of an RE in Fig. 21 by the pattern of Fig. 12 and connecting them appropriately, we have a configuration of ESPCA-01caef shown in Fig. 22 that simulates T_{parity}.

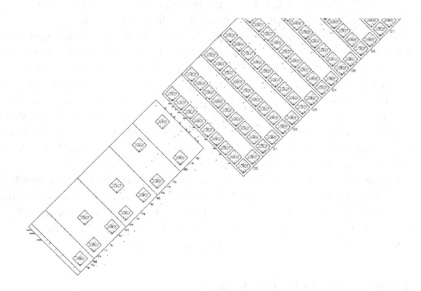

Fig. 22. RTM T_{parity} implemented in ESPCA-01caef simulated on Golly [10]. It realizes the circuit given in Fig. 21. Each small square is the pattern of an RE in Fig. 12

Similarly, we can obtain a configuration of ESPCA-02c5bf that simulates another RTM T_{power} that accepts the language $\{1^n \mid n \text{ is a power of } 2\}$ in Fig. 23.

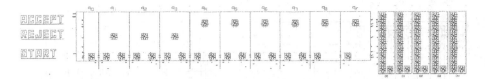

Fig. 23. RTM T_{power} implemented in ESPCA-02c5bf simulated on Golly [7]. Each small square is the pattern of an RE in Fig. 20

4.2 Intrinsic Universality of RCAs

We explain ESPCA-02c5bf is intrinsically universal with respect to the class of all reversible square PCAs (SPCAs). Note that SPCA is a superset of ESPCAs, where each part of a cell may have more than two states. As an example, we show here that ESPCA-02c5bf can simulate any evolution process of ESPCA-01caef. It is done by designing a pattern of a *metacell* in ESPCA-02c5bf that simulates one cell of ESPCA-01caef.

Let f be the local function of ESPCA-01caef (Fig. 6). As seen in Fig. 4(**b**), it is a mapping $f : (t, r, b, l) \mapsto (t', r', b', l')$. Figure 24 shows a circuit composed of REs and delay elements that computes the function f. It has three modules. They are the decoder, the state transition, and the decoder^{-1} modules.

The decoder module finds which one of the 16 cases of (t, r, b, l) occurs. By this, a decoded signal is obtained on one of the 16 lines. Using this information, the state transition module calculates the next state (t', r', b', l') from (t, r, b, l). Finally, in the decoder^{-1} module, 16 lines are reversibly merged into one. By this, the decoding signal is recycled and will be used in the next step. A more detailed description of the circuit is found in [11].

We then implement the circuit of Fig. 24 in the cellular space of ESPCA-02c5bf using the RE pattern given in Fig. 20. It is shown in Fig. 25 [10]. This pattern is the metacell for ESPCA-01caef, whose size is $10,000 \times 10,000$. Its center part contains a pattern that realizes the circuit given in Fig. 24. It simulates one step of ESPCA-01caef in 1,000,000 steps. Note that, in the emulator on Golly, besides the 16 states of ESPCA-02c5bf, several states are added to indicate the states of ESPCA-01caef, since it is very hard to recognize the state of ESPCA-01caef without such an indicator.

It is possible to simulate other reversible ESPCAs by changing the state transition module. Furthermore, by extending the three modules, any reversible SPCA can be simulated in ESPCA-02c5bf.

Using the RE pattern given in Fig. 12, we can implement a circuit of Fig. 24 in the cellular space of ESPCA-01caef. In this way, we see that ESPCA-01caef is also intrinsically universal.

As in the case of Turing universality, if a universal logic element is realizable in a CA, then it is, in principle, possible to compose a metacell pattern that simulates a cell of any given CA. However, the resulting pattern becomes very complex when the host CA is simple. By this reason, there have been a very

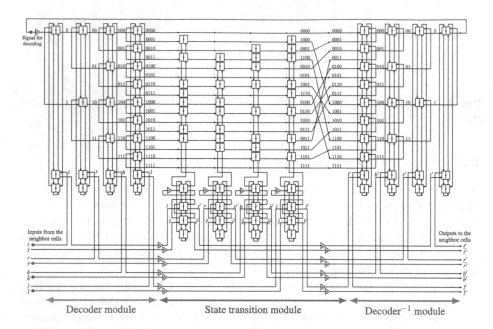

Fig. 24. Circuit composed of REs and delay elements that computes the local function of ESPCA-01caef [11]

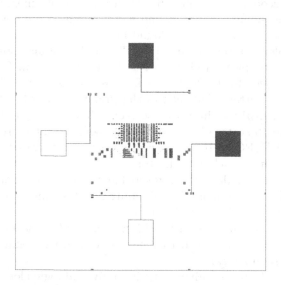

Fig. 25. Metacell in ESPCA-02c5bf that simulates one cell of ESPCA-01caef. Its size is $10,000 \times 10,000$. It is simulated on Golly [10]

few attempts to design a concrete metacell pattern in a CA. Such examples are found in the Game of Life CA (GoL), where several kinds of metacells have been created (see "Unit cell" of LifeWiki,[1] and patterns in the `HashLife/Metacell` folder of Golly [15]). In the above ESPCAs, an RE makes it easy to design a metacell systematically.

5 Concluding Remarks

We have investigated how an RE is implemented in very simple RCAs, and how it is used to show universality of the RCAs. Thanks to an RE, we can design complete configurations of RTMs and concrete patterns of metacells easily. Furthermore, using the CA simulator Golly [15], full computing processes of RTMs, and simulating processes of CAs by metacells become possible.

It is known that an RE is realizable also in ESPCA-01c5ef and ESPCA-02c5df [9]. Therefore, configurations of RTMs and metacells of reversible SPCAs can be designed in these ESPCAs likewise.

References

1. Bennett, C.H.: Logical reversibility of computation. IBM J. Res. Dev. **17**, 525–532 (1973). https://doi.org/10.1147/rd.176.0525
2. Fredkin, E., Toffoli, T.: Conservative logic. Int. J. Theoret. Phys. **21**, 219–253 (1982). https://doi.org/10.1007/BF01857727
3. Morita, Kenichi: A simple universal logic element and cellular automata for reversible computing. In: Margenstern, Maurice, Rogozhin, Yurii (eds.) MCU 2001. LNCS, vol. 2055, pp. 102–113. Springer, Heidelberg (2001). https://doi.org/10.1007/3-540-45132-3_6
4. Morita, K.: A new universal logic element for reversible computing. In: Martin-Vide, C., Mitrana, V. (eds.) Grammars and Automata for String Processing, pp. 285–294. Taylor & Francis, London (2003). https://doi.org/10.1201/9780203009642
5. Morita, K.: Theory of Reversible Computing. Springer, Tokyo (2017). https://doi.org/10.1007/978-4-431-56606-9
6. Morita, K.: Computing in a simple reversible and conservative cellular automaton. In: Das, S., Martinez, G.J. (eds.) ASCAT 2022. AISC, vol. 1425, pp. 3–16. Springer, Singapore (2022). https://doi.org/10.1007/978-981-19-0542-1_1
7. Morita, K.: Data set for simulating universal reversible elementary square partitioned cellular automata on Golly. Hiroshima University Institutional Repository (2023). https://ir.lib.hiroshima-u.ac.jp/00053792
8. Morita, K.: Making reversible computing machines in a reversible cellular space. Bull. EATCS **140**, 41–77 (2023). www.eatcs.org/index.php/on-line-issues
9. Morita, K.: Composing a rotary element in simple reversible cellular automata to make reversible computers. In: Adamatzky, A., Martinez, G.J., Sirakoulis, G.Ch. (eds.) Advances in Cellular Automata Springer (to appear)
10. Morita, K.: Data set for the Reversible World of Cellular Automata. Hiroshima University Institutional Repository (to appear). https://ir.lib.hiroshima-u.ac.jp/

---

[1] LifeWiki: The wiki for Conway's Game of Life, https://www.conwaylife.com/wiki/.

11. Morita, K.: Reversible World of Cellular Automata. World Scientific, Singapore (to appear). https://doi.org/10.1142/13516
12. Morita, K., Ueno, S.: Computation-universal models of two-dimensional 16-state reversible cellular automata. IEICE Trans. Inf. Syst. **E75-D**, 141–147 (1992). https://ir.lib.hiroshima-u.ac.jp/00048451
13. Ollinger, N.: Universalities in cellular automata. In: Rozenberg, G., Bäck, T., Kok, J.N. (eds.) Handbook of Natural Computing, pp. 189–229. Springer, Heidelberg (2012). https://doi.org/10.1007/978-3-540-92910-9_6
14. Toffoli, Tommaso: Reversible computing. In: de Bakker, Jaco, van Leeuwen, Jan (eds.) ICALP 1980. LNCS, vol. 85, pp. 632–644. Springer, Heidelberg (1980). https://doi.org/10.1007/3-540-10003-2_104
15. Trevorrow, A., Rokicki, T., Hutton, T., et al.: Golly: an open source, cross-platform application for exploring Conway's Game of Life and other cellular automata (2005). https://golly.sourceforge.net/

Effect of Delay Sensitivity in Life and Extended Life

Souvik Roy[✉]

School of Engineering and Applied Science, Ahmedabad University, Ahmedabad, Gujarat, India
souvik.roy@ahduni.edu.in,svkr89@gmail.com

Abstract. This paper shows the dynamics of Game of Life (Life) under delay-sensitive updating scheme where, during information sharing, neighbouring cells are associated with delay and probabilistic loss of information perturbation. Here, we explore the possibilities of continuous and abrupt change in phase during evolution of delay-sensitive Life. Next, we analyse the potential of micro-configurations (including oscillating, moving, stable micro-configurations) under delay-sensitive Life. Moreover, to understand the richness of Life, we observe the dynamics of extended Life rules and Life-like rules (both low and high density) under delay-sensitive environment.

Keywords: Game of Life · Life-like rules · Delay · Probabilistic loss of Information · Phase Transition

1 Introduction

In 1960's, the journey of cellular automata (CAs) was initiated by John von Neumann to capture the beauty of biological self-reproduction [16]. However, in front of scientific community, cellular automata was popularized by Conway's Game of Life (or, we can call simply Life) [3,7,12]. During the last four decades, CA researchers are searching for the answer to the key question – what is the trick or method for finding a simple rule like Life, which can be able to show wide range of surprising dynamics, including universal computation.

> *"I remember at some point being very curious about the origins of the Game of Life cellular automaton, and I spent several hours on the phone with John Conway – drilling and drilling and drilling, "Why did you come up with this? What are you doing?" And he kept on explaining it was a game, it was this, it was that. Eventually, I think I wore him down. And he explained that the real point was he had recently been hired as a professor of logic, which was not his primary field. And he wanted to do something interesting in the field of logic, and so he wanted to find a good enumeration of the recursive functions."* – Stephen Wolfram [22].

Moreover, another rich question, to what extent, Life rule can be able to capture the dynamics of real life (or, natural phenomenon). In order to understand the connection between Life and natural phenomenon in a better way, CA

M. Dalui et al. (Eds.): ASCAT 2024, CCIS 2021, pp. 17–30, 2024.
https://doi.org/10.1007/978-3-031-56943-2_2

researchers have explored the effect of perturbation (or noise) in Life. In early 1978, the resistance of Life against noise was first explored by Schulman and Seiden [21] where stochastic component 'temperature' acts as a noise (probability for birth or death) in the system. Though this stochastic model exhibits phase transition for critical value of temperature and is successfully predicted by mean-field analysis, the probabilities of the transition rules are not local (depending on global density, see [21]) which is the major drowback with this stochastic version of Life. In a similar direction, Adachi, Peper and Lee [2] introduces a new version of stochastic Life where the transition function is dependent on a temperature based continuous value expression. The stochastic Life of [2] shows $1/f$ and Lorentzian spectrum for respectively low and high temperature expression. To understand the effect of structural[1] and other perturbations, CA researchers have explored the effect of fully asynchronous [4], α-asynchronous [6,10,11], γ-asynchronous [10,11] updating schemes in Life. Nazim and Michel [10,11] have identified a phase transition from an 'inactive-sparse phase' to 'labyrinth phase' for changing value of α where each cell is updated with α probability. In a separate experiment, Blok and Bergersen [6] have also identified that Life shows a second-order phase transition under α-asynchronous updating scheme which also belongs to the directed percolation universality class [10]. Monetti and Albano [14,15] have also introduced stochastic rules in Life which shows first-order irreversible phase transition from steady state (non-zero density) to extinct phase (zero density). A good survey about Life and perturbation can found in [13].

Following the above literature, this paper explores the effect of delay sensitive [19,20] perturbation in Life which includes delay and probabilistic loss of information in terms of noise in the system. Traditionally, the updated state information of a cell (say, at time t) is immediately available to the neighbouring cell (at time $t + 1$), i.e. delay is not considered in the cellular system. On the other hand, a cell shares its state information with its neighbours deterministically, i.e. information loss is also not considered in the traditional system. However, these constraints are not true for natural systems, such as biological, physical, social systems. Here, in the delay sensitive system, the updated state information of a cell may not be available to its neighbours at the immediate time step. That is, a cell shares its state information with some delay which is also non-uniform in nature. Moreover, a cell shares its state information with its neighbours probabilistically, that is, information loss is considered in the delay sensitive model. Under such circumstances, this paper investigates the following key research questions – (1) Is Life sensitive to the effect of probabilistic loss of information? (2) What will be the effect of (non-uniform) delay in the Life dynamics? (3) Does Life show any surprising dynamics under the proposed (overall) delay sensitive system?

Next, to understand the richness of Life (that is, the starting question, what is there so exciting about Life?), we explore the dynamics of low density Life-like rules (Honey Life, Flock Life, Eight Life, High Life, Pedestrian Life

---

[1] Here, the term 'structural' indicates modification in the topological component interaction.

etc.), high density `Life` like rules (Drigh `Life`, Dry `Life` etc.) and extended `Life` rules under delay sensitive environment. To understand all these questions, next we introduce the proposed delay sensitive system.

2   The Delay-Sensitive Life and Experimental Protocol

Traditionally, `Life` evolves in a regular subset of $\mathbb{Z}^2$ with state set $S = \{0, 1\}$ where 0 stands for dead and 1 stands for alive. However, during the experimental study, we consider configuration of finite squares with $N \times N$ cells under periodic boundary condition, i.e. $\mathbb{Z}/N\mathbb{Z}$. Note that, `Life` shows completely different dynamics for other (specifically, open) boundary condition [5]. The neighbourhood dependency of `Life` follows Moore neighbourhood, i.e. self and eight nearest neighbours. Following are the local transition rule of `Life`.

1. A dead cell with exactly three live neighbours evolves to live state, i.e. birth rule. We denote by `B3` where `B` represents birth; and
2. A live cell with exactly two or three live neighbours will remain alive, i.e. survival rule. We denote by `S23` where `S` depicts survival.

To sum up, `Life` rule can be written as `B3/S23` In general, `Bp/Sq` is traditionally used for the naming of these 2-D outer-totalistic rules where p and q are the subsets that can contains digits from 0 to 8 to represent the number of live neighbours. Note that, we follow this naming approach to represent extended `Life` rules and `Life`-like rules.

Traditionally, `Life` follows synchronous updating scheme. In this paper, we consider following delay sensitive updating scheme.

– Every two neighbouring cell i, j, where $i \neq j$, are associated with non-negative integer function delay `D(i,j)`. And, `D(i,j) = D(j,i)`.
– `D(i,j)` $\geqslant 1$, that is, if `D(i,j)` $= k$ then the updated state information of cell i at time t is available to cell j after $t + k$ time steps, or vice versa. Note that, for traditional synchronous system `D(i,j) = 1`.
– The system follows non-uniform delay, that is `D(i,j)` may differ with `D(i1,j1)` where i, j and $i1, j1$ are neighbouring cells.

To sum up, the delay perturbation parameter $\mathcal{D} \in \mathbb{N}$ depicts the limit of maximum delay where every neighbouring pairs (cells) are associated with a random (uniform distribution) delay between 1 to $\mathcal{D}$.

– Moreover, when a cell shares its state information with it's neighbours, the information sharing is associated with a probabilistic loss of information rate ι, where $0 \leqslant \iota \leqslant 1$.

To implement this delay sensitivity, we redefine the cellular system framework, where each cell is associated with a view of its neighbour's state which changes depending on the arrival of the (new) state information of the neighbours. Therefore, the state set is associated with *actualstate* and *viewstate* of

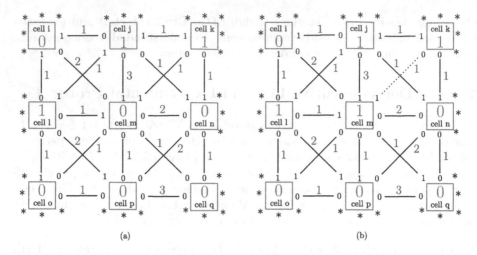

Fig. 1. (a) Delay sensitive Life at time t where each link associated with delay (in blue). The state of cells are marked with red. The viewstates of cells are in black; (b) The system at time $t + 1$. Dotted line depicts the probabilistic loss of information. (Color figure online)

the neighbours. Naturally, we need to sub-divide the local transition function into two parts - *state update step* and *information sharing step*. Let us consider, an instance of `Life` (see Fig. 1(a)) where state of the cells are marked with red within the box. Every pair of cells are associated with delay (in blue), for example D(j,m) = 3, D(i,m) = 2, D(p,m) = 1. Moreover, each cell is associated with a view (viewstate), about the states of neighbours. Cell m sees that state of cell j, k, l is 1 and state of cell i, n, o, p, q is 0, noted with black in the direction of neighbouring cells. '⋆' indicates the viewstates of neighbours which are not explicitly mentioned in this instance of `Life`. Now, following the `Life` rule, cell m in state 0 with three live neighbours moves to state 1 during the t time step. D(m,p) = 1, therefore, the information about the state change of cell m is visible to cell p at $t + 1$ time step, however, the information is not available to cell j at $t + 1$ time step because D(m,j) = 3, see Fig. 1(b). On the other hand, though D(m,k) = 1, the information about the state change of cell m is not available to cell k due to probabilistic loss of information (marked with dotted line) at $t + 1$ time step, see Fig. 1(b). Hence, to cell k, state of cell m is 0, though the (actual) state of cell m is 1 during that time. In our early work, we have formally defined this extended framework, see [19] for details. To sum up, in this experimental setup, the information loss perturbation follows $0 \leqslant \iota \leqslant 1$, and, the delay perturbation parameter $\mathcal{D} \in \mathbb{N}$ depicts the limit of maximum delay. In this experiment, $\mathcal{D} \in [1, 4]$. Note that, in this proposed model, delay for a neighbouring pair remains constant over time.

Here, we follow qualitative and quantitative experimental approach to understand the effect of noise in `Life`. In the first qualitative approach, we start with

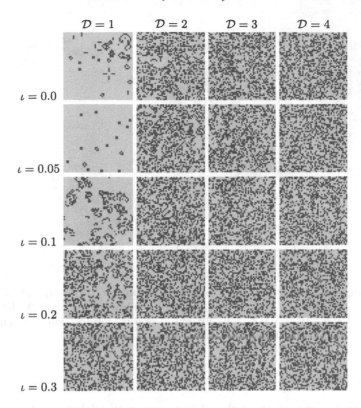

Fig. 2. Life configuration during evolution after $t_{transient} = 2000$ time steps where the initial configuration is associated with $d_{ini} = 0.5$. Here, the CA size is 50×50. The alive and dead states are marked by blue and cyan colour respectively. This convention is kept (if not mentioned). (Color figure online)

50×50 random configuration starting with a fixed density (no of alive cell). Hereafter, during the evolution of the system, we need to observe the space-time diagrams of the system. Following this approach, we can be able to provide a visual comparison between traditional **Life** and delay sensitive **Life**. Secondly, in the formal quantitative method, we study the (steady-state) density of the configuration which can be defined as $d_x = \frac{\#_{alive}x}{|x|}$. Here, $\#_{alive}x$ depicts the number of alive cells in the configuration x and $|x|$ is the size of the configuration space (here, 50×50). In this approach, we evolve the **Life** for a transient time period, say $t_{transient}$, starting from an initial configuration with density d_{ini}. In this experiment, $t_{transient} = 2000$ time steps. Hereafter, we calculate the average density of **Life** for a sampling period of time, say $t_{sampling}$, which is denoted by d_{avg}. Here, $t_{sampling} = 100$ time steps. The work of [10,11,19] also follows similar quantitative sampling algorithm. Therefore, $d_{avg}(d_{ini}, \iota, \mathcal{D})$ indicates the steady-state density under the proposed delay sensitive updating scheme. In the next section, we report the effect of **Life** under this proposed updating scheme.

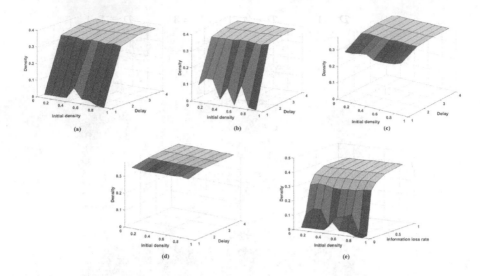

Fig. 3. Sampling surfaces according to density parameter - (a), (b), (c), (d): initial density-delay-density, i.e. $d_{ini} - \mathcal{D} - d_{avg}$, (x-y-z), where in (a) $\iota = 0.0$, (b) $\iota = 0.1$, (c) $\iota = 0.2$, and (d) $\iota = 0.3$; and (d): initial density-information loss-density, i.e. $d_{ini} - \iota - d_{avg}$, (x-y-z), where $\mathcal{D} = 1$.

3 Dynamics of Delay Sensitive Life

This section reports the dynamics of Life under the delay sensitive updating scheme following the qualitative and quantitative experimental protocol. Let us first focus on only probabilistic loss of information perturbation. The left-most column of Fig. 2 shows the change in space-time diagram of Life after $t_{transient} = 2000$ time step. Observe that, Life shows *inactive-sparse* phase for small value of information loss rate, i.e. $\iota = 0.0$, 0.05, 0.1, see Fig. 2. However, *active-dense* phase is observed for relative large value of ι, where $\iota = 0.2$ and 0.3. Here, $d_{avg} \geqslant 0.2$ is considered as *active-dense* dynamics. Figure 3(e) depicts the quantitative sample surface $(d_{ini}\text{-}\iota\text{-}d_{avg})$ where a phase transition (from 'inactive-sparse' to 'active-dense') is observed for increasing value of information loss perturbation rate. Note that, for different non-classical variant of CA [10,11], appearance of phase transition is one of the rich phenomenon in terms of statistical physics. Figure 3(e) clearly shows that this phase transition phenomenon is following second order transition or continuous in nature. In a similar experiment with 'small' scale link removal, Fatés [10,11] reported the dynamics of Life. However, the phase transition behaviour was not noted by [10,11] due to short range of link removal experimental setup.

Next, we consider only delay perturbation where information loss rate ($\iota = 0.0$) remains static. Figure 2 (topmost row) shows the qualitative space-time diagrams of Life for $\mathcal{D} \in [1, 4]$, whereas Fig. 3(a) depicts the quantitative sample surface $(d_{ini}\text{-}\mathcal{D}\text{-}d_{avg})$ where $\iota = 0.0$. Figure 2 shows the *inactive-sparse* phase for $\mathcal{D} = 1$ (classical Life) and *active-dense* phase for $\mathcal{D} \in [2, 4]$. However,

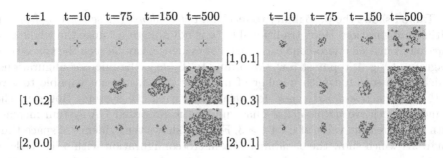

Fig. 4. Life configurations for 50×50 after $t = 10$, 75, 150 and 500 time steps, starting from a initial configuration with 9 live cells ($t = 1$). The top most left row depicts the situation for classical Life.

here, this phase change is different from probabilistic loss of perturbation. Under delay perturbation, the phase transition is abrupt, not continuous in nature. Can be observed that, for $\mathcal{D} \in [2, 4]$, the dynamics remain same in space-time diagram (see Fig. 2, topmost row) and sample surface (see Fig. 3(a)). Now, if we consider the proposed overall delay sensitive system, i.e. delay and information loss perturbation together, we can easily predict the dynamics from the individual behaviour. Figure 2 (second and third row respectively) shows the dynamics for $\iota = 0.05$ and $\iota = 0.1$. Here also, the abrupt change in phase is visible for increasing value of $\mathcal{D}$ parameter. Figure 3(b) shows the evidence of sample surface (d_{ini}-$\mathcal{D}$-d_{avg}) where $\iota = 0.1$. However, Life does not show any change for increasing value of $\mathcal{D} \in [1, 4]$ when $\iota = 0.2$ and $\iota = 0.3$. As an evidence, can be seen the flat sampling surface of Fig. 3(c) and Fig. 3(d) where the static information loss perturbation rate is 0.2 and 0.3 respectively.

4   Dynamics of Micro-configurations

This section shows the effect of delay sensitivity in Life following two other experimental setups.

In the first experiment, we perform the following well established experimental study [10, 11] to analyse the potential of the micro-configurations (i.e. a very localized part of initial configurations considering only alive states and nearby dead states.) of Life: (a) the experiment starts with an initial configuration having i number of live cells, where $i = 1, 2, \cdots, 9$. Here, the arrangement of live cells can be viewed as a 3×3 configuration (complete or semi-complete part of 3×3 grid). (b) Therefore, we study the evolution of Life until it reaches to a fixed point or *active-dense* phase. Here again, we assume that the minimum density limit is 0.2 for achieving *active-dense* phase. Now, we repeat this experiment 100 times for each value of i (where $i = 1, 2, \cdots, 9$). Therefore, the study analyses the expressiveness of Life starting from a very localized part of the lattice. According to [11], we can call these initial configurations as *germs* which may have the capability to establish an *active-dense* phase.

Table 1 shows the probability to reach an *active-dense* phase starting from an initial configuration. Here, traditional Life is not able to colonize the whole grid starting from 3×3 micro-configurations. For evidence, can be seen the space-time diagram of Fig. 4, topmost row. According to Table 1, the micro-configurations, which are associated with number of live cells $i < 3$, are not capable to give birth of an *active-dense* phase for delay-sensitive Life system. However, the chances to achieve *active-dense* phase under the proposed CA system increase with the increasing value of i, for $i \geqslant 3$. Figure 4 shows the evidence of space-time diagrams which show the capability of a micro-configuration with 9 live cells to establish *active-dense* phase, here, $\mathcal{D} = 1, \iota = 0.1$; $\mathcal{D} = 1, \iota = 0.2$; $\mathcal{D} = 1, \iota = 0.3$; $\mathcal{D} = 2, \iota = 0.0$ and $\mathcal{D} = 2, \iota = 0.1$.

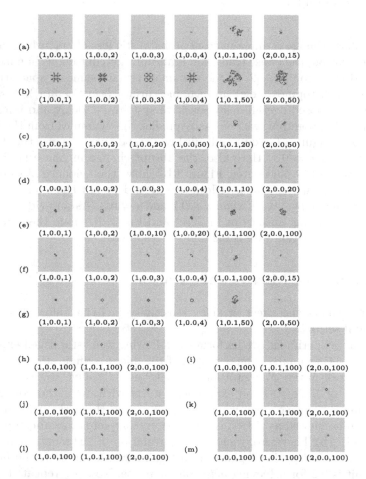

Fig. 5. Dynamics of Life micro-configurations (a) Blinker; (b) Pulsar; (c) Glider; (d) Toad; (e) Spaceship; (f) Beacon; (g) Largeblock; (h) Beehive; (i) Boat; (j) Loaf; (k) Ring; (l) Ship; and (m) Tub under delay-sensitive updating scheme.

Table 1. The probability for a micro configuration with i number of live cells (where, $i = 1, 2, \cdots, 9$) to establish an active-dense phase.

i	1	2	3	4	5	6	7	8	9	i	1	2	3	4	5	6	7	8	9
$D=1, \iota=0.0$	.00	.00	.00	.00	.00	.00	.00	.00	.00	$D=1, \iota=0.1$	.00	.00	.00	.00	.00	.00	.00	.00	.00
$D=1, \iota=0.2$	.00	.00	.05	.13	.24	.33	.37	.39	.51	$D=1, \iota=0.3$	.00	.00	.07	.17	.28	.37	.41	.44	.56
$D=2, \iota=0.0$	.00	.00	.03	.09	.19	.21	.39	.43	.52	$D=2, \iota=0.1$	.00	.00	.05	.13	.24	.31	.40	.44	.55

In the second experiment, we analyze the potential of following well-known micro-configurations of Life under the proposed delay-sensitive updating scheme: (i) oscillating patterns (oscillate in a fixed position of the grid)[2] (ii) moving patterns (moves across the grid)[3] and (iii) stable patterns (remain fixed during the evolution or fixed point)[4]. In this experiment, we start with the initial configuration of these patterns to distinguish between classical Life and delay-sensitive Life. Figure 5 reports the experimental results where $(\mathcal{D}, \iota, t)$ denotes the space-time diagram after t time steps. Observe that, all the oscillating and moving patterns are destroyed for little perturbation, $\mathcal{D} = 2$ or $\iota = 0.1$. Here, oscillating pulsar shows *active-dense* during the evolution. However, all stable patterns show resistance against delay-sensitive perturbation. These patterns (stable) remain in fixed point for $\mathcal{D} \in [1, 4]$ and $\iota \in [0, 1]$, see Fig. 5.

5   Dynamics of Extended Life

According to the study, Life shows continuous and abrupt phase transition respectively for probabilistic loss of information and delay perturbation. In this section, we explore the dynamics of Life-like rule to understand the richness of Life. In a early work, Torre and Mártin [8] have reported the behaviour of Life-like rules under synchronous environment. A good survey about the dynamics of Life-like rules can found in [9]. Recently, Peña and Sayama [17] have explored the complexity of Life-like rules using the notion of conditional entropy. According to [17], in comparison with Life and Life-like rules, Life is associated with low density and high information content throughout the evolution. On the contrary, both Life and extended Life rules B1234567/S56, B3S12, B3456S3456, B3456/S6 show phase transition dynamics under α-asynchronous updating scheme [10]. Similarly, Kaleidoscope of Life also shows a second-order phase transition [1] under asynchronous updating scheme.

Following the literature, this study consider following low and high density Life-like rules and extended Life rules and under delay sensitive environment:

First, we explore low density Life-like rules under delay sensitive environment which include B38/S238 (Honey Life), B368/S238 (Low Death Life), B3/S238 (Eight Life), B38/S23 (Pedestrian Life), B3/S12 (Flock Life),

---

[2] a) blinker, (b) pulsar, (d) toad, (f) beacon, (g) largeblock.
[3] (c) glider, (e) spaceship.
[4] (h) beehive, (i) boat, (j) loaf, (k) ring, (l) ship, (m) tub.

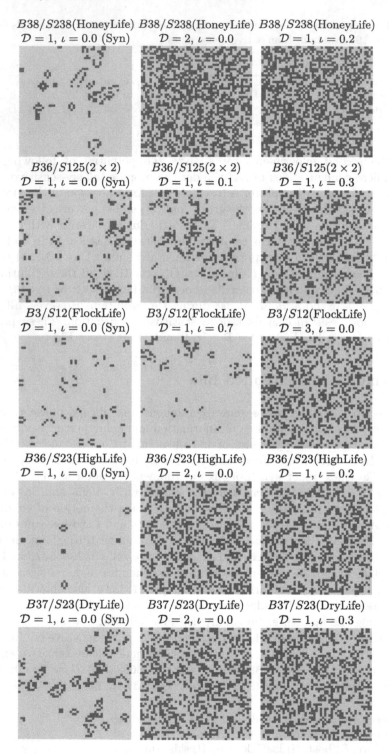

Fig. 6. Space-time diagrams of Life-like rules after transient time steps.

Low density Life-like rules	B38/S238, B368/S238, B3/S238, B38/S23, B3/S12, B36/S23, B36/S125
High density Life-like rules	B367/S23, B3578/S238, B3578/S23, B356/S23, B37/S23
Extended Life rules	B1234567/S56, B3456/S3456, B12345/S12345, B1234/S1234, B3456/S6

B36/S23 (High Life), B36/S125 (2 × 2 Life). Figure 6 shows some space-time diagrams for Honey Life, 2 × 2 Life, Flock Life, High Life under the proposed delay sensitive updating scheme. Remark that, all the above mentioned low density Life-like rules show abrupt phase transition for delay perturbation. Following the dynamics of Life, all these rules show *inactive-sparse* phase for $\mathcal{D} = 1$ and *active-dense* phase for $\mathcal{D} \in [2, 4]$. On the other hand, most of the low density Life-like rules depict a second order phase transition for probabilistic loss of information perturbation where the critical value of phase transition is $\iota_c \approx 0.2$. Note that, we have observed the same ι_c for Life under probabilistic loss of information perturbation. In fact, under α-asynchronous updating scheme, the work of [10] have also reported the similar critical value for phase transition. The reason behind this signature behaviour is still open to us. On a contrary, Flock Life (B3/S12) shows a continuous phase transition for very high value of probabilistic loss of information where $\iota_c \approx 0.8$, for evidence see Fig. 6 ($\iota = 0.7$). Similarly, we also explore high-density Life-like rules – B367/S23 (Drigh Life), B3578/S238, B3578/S23, B356/S23, B37/S23 (Dry Life) under delay sensitive environment. Again, delay perturbation reflects abrupt change in phase for all these high density Life-like rules, for evidence, see Fig. 6 for Dry Life. As an interesting result, B3578/S238, B3578/S23, B356/S23 show continuous phase transition for very early probabilistic loss of information rate where $\iota_c \approx 0.05$. In this context, Drigh Life and Dry Life shows similarity with Life, i.e. $\iota_c \approx 0.2$. Note that, many of these rules contain Life as subset within them. Finally, we can remark that phase transition (both abrupt and continuous) under delay sensitive environment is a general phenomenon for Life and Life-like rules.

Moreover, some extended Life rules also depict interesting dynamics. For example, B1234567/S56 shows *"dense labyrinth"* patterns for traditional synchronous system, however, those labyrinth patterns are destroyed with the effect of delay sensitivity, see Fig. 7. In the literature, Fatès [10] have observed the same under α-asynchronous perturbation. On the other hand, B3456/S3456 converges for little probabilistic loss of information, and the convergence configurations show wider variety of box pattern, however, this convergence is not reflected for only delay perturbation. The same behaviour is reflected by B12345/S12345. Again, note that, α-asynchronism depicts the same dynamics [10] for B3456/S3456. In a similar direction, for B1234/S1234, one can observe 'big flashing islands' pattern under traditional updating scheme, however, for both delay and information loss perturbation, it shows dense-labyrinth pattern.

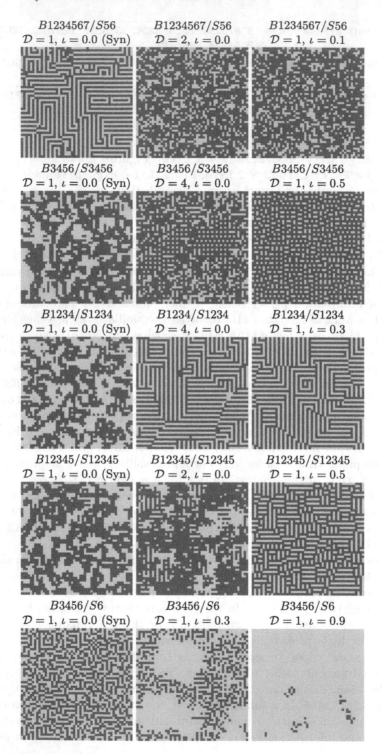

Fig. 7. Space-time diagrams of Extended Life rules after transient time steps.

According to Regnault et al. [18], B1234/S1234 also depicts labyrinth pattern under fully asynchronous updating scheme. In a different direction, B3456/S6 depicts a phase transition from steady state (non-zero density) to extinct (zero density) with the effect of probabilistic loss of information perturbation, see Fig. 7. Note that, the same phase transition is also observed for α-asynchronous perturbation [10]. Moreover, Monetti and Albano [14,15] have also introduced stochastic rules in Life which shows first-order irreversible phase transition from steady state (non-zero density) to extinct phase (zero density). However, the reason behind similar behaviour under both of these perturbations (delay sensitivity and α-asynchronism) is still open to us.

6 Conclusion

To sum up, Life shows continuous phase transition from *inactive-sparse* to *active-dense* for probabilistic loss of information perturbation. However, the effect of delay perturbation in Life is abrupt (abrupt change in phase). This abrupt change is also visible for system considering both of the perturbations where information loss perturbation is relatively little. On the other hand, delay sensitive Life can be able to *colonize* the whole grid starting from small micro-configuration (in fact, starting from three alive cells). On a contrary, stable micro-configurations show solid resistance against perturbation.

On the other hand, low and high density Life-like rules (Honey Life, Low Death, Eight Life, Pedestrian Life, Flock LifeHigh Life, 2 × 2 Life, Drigh Life, Dry Life) also show this continuous and abrupt phase transition for the effect of probabilistic loss of information and delay perturbation respectively. Moreover, Life and Life-like rules show similar critical value for phase transition. It indicates that this phase transition is a common phenomenon for Life and Life-like rules.

Another important observation, delay-sensitive and α-asynchronous perturbation show similar dynamics for extended Life rules B1234567/S56, B3456/S3456, B3456/S6. In fact, B1234/S1234 shows same signature behaviour for delay-sensitive and fully-asynchronous perturbation. However, the reason behind these signature behaviour is still open to us after this first experimental study.

References

1. Adachi, S., Lee, J., Peper, F., Umeo, H.: Kaleidoscope of life: a 24-neighbourhood outer-totalistic cellular automaton. Physica D **237**(6), 800–817 (2008)
2. Adachi, S., Peper, F., Lee, J.: The game of life at finite temperature. Physica D **198**(3), 182–196 (2004)
3. Berlekamp, E.R., Conway, J.H., Guy, R.K.: Winning Ways for Your Mathematical Plays, vol. 2. Academic Press, London (1984)
4. Bersini, H., Detours, V.: Asynchrony induces stability in cellular automata based models. In: Artificial Life IV: Proceedings of the Fourth International Workshop on the Synthesis and Simulation of Living Systems, pp. 382–387. The MIT Press (1994)

5. Blok, H.J., Bergersen, B.: Effect of boundary conditions on scaling in the "game of life". Phys. Rev. E **55**, 6249–6252 (1997)
6. Blok, H.J., Bergersen, B.: Synchronous versus asynchronous updating in the "game of life". Phys. Rev. E **59**, 3876–3879 (1999)
7. Das, S., Roy, S., Bhattacharjee, K.: The Mathematical Artist: A Tribute To John Horton Conway. Emergence, Complexity and Computation, Springer, Cham (2022). https://doi.org/10.1007/978-3-031-03986-7
8. de la Torre, A.C., Mártin, H.O.: A survey of cellular automata like the "game of life". Physica A Stat. Mech. Appl. **240**(3), 560–570 (1997)
9. Eppstein, D.: Growth and Decay in Life-Like Cellular Automata, pp. 71–97. Springer, London (2010). https://doi.org/10.1007/978-1-84996-217-9_6
10. Fatès, N.: Does Life Resist Asynchrony?, pp. 257–274. Springer, London (2010). https://doi.org/10.1007/978-1-84996-217-9_14
11. Fatès, N., Morvan, M.: Perturbing the topology of the game of life increases its robustness to asynchrony. In: Sloot, P.M.A., Chopard, B., Hoekstra, A.G. (eds.) ACRI 2004. LNCS, vol. 3305, pp. 111–120. Springer, Heidelberg (2004). https://doi.org/10.1007/978-3-540-30479-1_12
12. Gardner, M.: Mathematical games: the fantastic combinations of John Conway's new solitaire game "life". Sci. Am. **223**(4), 120–123 (1970)
13. Martínez, G.J., Adamatzky, A., Seck-Tuoh-Mora, J.C.: Some Notes About the Game of Life Cellular Automaton, pp. 93–104. Springer, Cham (2022). https://doi.org/10.1007/978-3-031-03986-7_4
14. Monetti, R.A.: First-order irreversible phase transitions in a nonequilibrium system: mean-field analysis and simulation results. Phys. Rev. E **65**, 016103 (2001)
15. Monetti, R.A., Albano, E.V.: Critical edge between frozen extinction and chaotic life. Phys. Rev. E **52**, 5825–5831 (1995)
16. Neumann, J.V.: Theory of Self-Reproducing Automata. University of Illinois Press, Illinois (1966)
17. Peña, E., Sayama, H.: Life worth mentioning: complexity in life-like cellular automata. Artif. Life **27**(2), 105–112 (2021)
18. Regnault, D., Schabanel, N., Thierry, É.: On the analysis of "Simple" 2D stochastic cellular automata. In: Martín-Vide, C., Otto, F., Fernau, H. (eds.) LATA 2008. LNCS, vol. 5196, pp. 452–463. Springer, Heidelberg (2008). https://doi.org/10.1007/978-3-540-88282-4_41
19. Roy, S.: A study on delay-sensitive cellular automata. Phys. A **515**, 600–616 (2019)
20. Roy, S., Das, S., Mukherjee, A.: Elementary cellular automata along with delay sensitivity can model communal riot dynamics. Complex Syst. **31**(3), 341–361 (2022)
21. Schulman, L.S., Seiden, P.E.: Statistical mechanics of a dynamical system based on Conway's game of life. J. Stat. Phys. **19**(3), 293–314 (1978)
22. Wolfram, S.: Two Different Directions: John Conway and Stephen Wolfram, pp. 21–71. Springer, Cham (2022). https://doi.org/10.1007/978-3-031-03986-7_2

Analysis of Traffic Conflicts in Signalized Intersections Using Cellular Automata

R. Krishna Kumari[1](✉) ⓘ, K. Janaki[2] ⓘ, and R. Arulprakasam[3](✉) ⓘ

[1] Career Development Centre, College of Engineering and Technology,
SRM Institute of Science and Technology, Kattankulathur, Chennai 603203,
Tamil Nadu, India
krishrengan@gmail.com
[2] Department of Mathematics, Saveetha Engineering College, Saveetha Nagar,
Chennai 602105, Tamil Nadu, India
[3] Department of Mathematics, College of Engineering and Technology,
SRM Institute of Science and Technology, Kattankulathur, Chennai 603 203,
Tamil Nadu, India
r.aruljeeva@gmail.com

Abstract. Conflicts between vehicles and pedestrians play a vital role in urban areas. In this paper, cellular automata is used as a simulation model to study traffic conflicts between vehicles and pedestrians in three way signalized intersections. Three vehicular streams and two pedestrian streams are considered for the study. The signal time cycle is fixed as 90 s which includes 2 signal phases that are unique for both vehicles and pedestrians. The simulation process indicates that cellular automata can be used as an effective method to analyse congestion and conflicts between pedestrian-vehicular streams.

Keywords: Cellular automata · Road intersection · Pedestrian-Vehicular conflicts

AMS Subject Classification: 68Q80, 68Q85

1 Introduction

In India, a significant portion of the mixed traffic flow at urban road intersections is made up of pedestrians [1,5,6,16]. However, they are at a higher risk of suffering injuries and even dying in a pedestrian-vehicle collision, making them more vulnerable road users. Therefore, in order to create a safe, effective traffic operation environment, it is crucial to investigate the mechanisms of traffic conflicts between pedestrians and vehicles at intersections [7,14]. Prior research on traffic conflicts in India mainly concentrated on the following areas:

1. The identification of traffic conflicts and how to assess their seriousness;
2. The prediction of traffic conflicts and how they relate to traffic accidents;
3. The use of technology related to traffic conflicts in safety evaluation, etc.

M. Dalui et al. (Eds.): ASCAT 2024, CCIS 2021, pp. 31–44, 2024.
https://doi.org/10.1007/978-3-031-56943-2_3

Nearly all of these studies' findings require data on traffic conflict as a prerequisite. However, the collection of data primarily depends on the investigators' subjective assessment, which cannot ensure consistency between various observations. Furthermore, more studies of traffic conflict focus on the vehicle flow rather than the interaction between pedestrians and vehicles as a result of the rapid growth in urban vehicles and the worsening of traffic congestion.

In late 1940, cellular automata (CA) [13] was first introduced by Von Neumann and Ulam as a model for investigating the behavior of complex systems [3,10]. Diversity and simple behavior of CA has made it suitable for various sciences - from ecology and biology to computer sciences, mathematics, and physics. CA is a mathematical model that can be used for calculation and simulation of systems. CA refers to simple discrete systems that can exhibit complex behaviors and computations via simple and local rules. In other word, adjacent neighborhoods affect the behavior of each cell while farther neighborhoods have no such effect on the cell behavior. In recent years cellular automata is used as a simulation model because of their simplicity and flexibility. CA has been widely used in the microscopic simulation of urban traffic. CA is uniquely flexible and adaptable in various situations due to the discreteness of the time, space, and status of the traffic flow elements, which is appropriate to describe the variety and complexity of pedestrian flow problems. Consequently, CA has a significant advantage in modelling and detecting conflicts in the study of traffic conflict and in describing the behaviour choices of pedestrians and vehicles. The decision to employ cellular automata for studying pedestrian movements reflects the method's advantageous characteristics. Pedestrian simulation poses greater challenges compared to vehicle simulation due to the inherent complexity and randomness associated with pedestrian movements. Despite this complexity, the cellular automata simulation method proves effective in handling the intricacies and uncertainties inherent in pedestrian movement processes [4,15]. Rather than relying on elaborate mathematical models, the approach leverages traffic behavior rules to appropriately describe and simulate pedestrian interactions.

The primary objectives of this simulation study are to quantitatively assess crosswalk capacity, determine the number of traffic conflicts arising between pedestrians and vehicles, and analyze pedestrian delay resulting from these conflicts at signalized intersections [2,9,11]. The study further aims to explore the relationships between these variables and two key factors: pedestrian signal timing and crosswalk width [8,12]. In order to provide a theoretical foundation for the assessment of pedestrian safety, this paper investigates the method of rule description for pedestrian and vehicle behaviours as well as a cellular automaton model to analyse the traffic conflicts between them at signalised intersections.

2   Preliminaries

In this section, some basic concepts are recalled.

Definition 1. *A Cellular automata (CA) is a discrete model consisting of a regular grids of cells such that the grid can exist in any finite number of dimensions.*

*Each cell changes state as a function of time accordingly to a defined set of rules
that includes the states of neighbouring cells. Formally it can be expressed as a
5-tuple $C = (R, M, A, f, a_0)$, where*

- *R is a regular grid of cells,*
- *A is a set of finite states,*
- *a_0 is initial configuration, $(a_0 \in A)$,*
- *M is a finite set of neighborhood indices for all $r \in R, c \in M : r + c \in R$, and*
- *$f : A^n \to A$ is transition function.*

Definition 2. *Road intersections are the junctions or the meeting point of two
or more roads.*

3   Cellular Automata Model

The three way intersection considered in this model is a signalized intersection
with a fixed vehicular-pedestrian signal time cycle of 90 seconds. Totally five
streams are considered of which three are vehicular streams and two are pedestrian streams (see Fig. 1). The model used in this study simulates two crosswalks
at a signalised intersection with a 30 second green light time and a 60 second red
light time for the pedestrian signal. The road segment leading to the intersection
is designed with two-way two vehicle lanes each measuring 12 meters in width.
The crosswalk has a length of 24 meters and a width of 3 meters. Additionally,
there are two pedestrian waiting areas on either side of the crosswalk, each of
which has room for up to 100 people. The crosswalk in this model is divided
into cells that are each $0.5\,\text{m}^2$ square. There is a single pedestrian in each cell,
or the cell is empty. The pedestrian's speeds can range from 2, 3, 4, 5 or 6 cells
per second, which corresponds to 1 m/s, 1.5 m/s, 2 m/s, 2.5 m/s, and 3 m/s,
respectively, in terms of their actual speeds. At the crosswalk, each vehicle is
assumed to move at a speed of 6 m/s and occupy a space of 6×5. The open
boundary condition is used, and each simulation time step's span is set to one
second. In Fig. 2,

- σ_1 denotes the straight moving vehicle stream towards the right direction.
- σ_2 denotes the straight moving vehicle stream towards the left direction.
- σ_3 denotes the left turning vehicle stream.
- σ_4 denotes the bidirectional Pedestrian stream in the one way road.
- σ_5 denotes the bidirectional Pedestrian stream in the two way road.

Signal groups 1 and 2 are shown in Figs. 3 and 4.

Fig. 1. Real Life Three Way Intersection with Uncontrolled Traffic

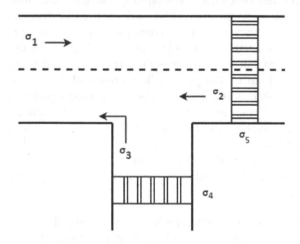

Fig. 2. Three way Intersection

3.1   Pedestrian Rules

1. The arrival rate of the bidirectional pedestrian flow obeys the poisson distribution as shown in

$$P_m = \frac{(\lambda_p t)^m}{m!} e^{\lambda t}$$

where, P_m denotes the probability of reaching m pedestrians, during the interval. λ_p denotes the average pedestrian arrival rate (capita/second). t denotes each of the time interval counted.

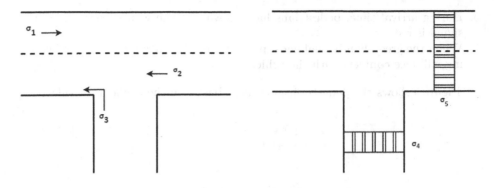

Fig. 3. Signal group 1 (60 s)   **Fig. 4.** Signal group 2 (30 s)

2. A bi-directional microscopic model should consider the following fundamental elements as rules for pedestrian. (see Fig. 2)
   (i) Side stepping (lane changing)
   (ii) Forward movement (braking, acceleration)
   (iii) Conflict mitigation (deadlock avoidance)
3. Pedestrians give importance to forward movement to the front cell as their destination. They change lanes to the left or right cell when he/she is blocked in front provided the neighbour cell must be empty.
4. Let $V_{i,j}$ denote the pedestrian's velocity and $x_{i,j}$ denote the location of pedestrian. Let $d_{i,j}$ denote the gap in the form of empty cells between the pedestrian and his/her nearest vertically front pedestrian. Velocity of pedestrian in crosswalk is calculated as

$$V_{i,j} = min\{d_{i,j}, v_{i,j}\}$$

The initial velocity of pedestrian going into the crosswalk is assigned according to field observation as follows :

$$V_{i,j} = \begin{cases} 2 & \text{if } 0.2 \geq p \geq 0 \\ 3 & \text{if } 0.5 \geq p \geq 0.2 \\ 4 & \text{if } 0.7 \geq p \geq 0.5 \\ 5 & \text{if } 0.9 \geq p \geq 0.7 \\ 6 & \text{if } p > 0.9. \end{cases}$$

where, p is a random number between 0 and 1.
5. During peak hours, many pedestrians complete for one empty cell and this leads to pedestrian-pedestrian conflicts. During that situation, the system selects one pedestrian at random to occupy that desired cell with an equal probability.

6. During arrival time, pedestrians have to wait at the waiting area when the signal is red.
7. When the green light signal ends, pedestrians who have not completed crossing will face conflicts with the vehicles.

Figure 5 shows the pedestrian movement based on the elementary rules

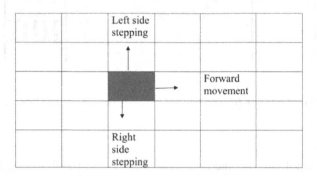

Fig. 5. Pedestrian movement based on the elementary rules

3.2   Vehicle Rules

1) The arrival rate of vehicular flow also obeys the poisson distribution similar to that of pedestrian.
2) When the pedestrian light is green and if the rate of flow of pedestrian is low or nil, vehicles are not allowed. Pass through that lane since the considered model is a signalised model.

3.3   Areas of Conflict

In this paper, the conflicts between pedestrians and vehicles exist in two areas.

(1) Conflicts between left turning vehicles σ_3 and pedestrian stream σ_4 occurs during the signal 2. Although green light signal is for σ_4, assuming that left turn on red is allowed at the approach leads to conflicts.
(2) Conflicts between straight moving vehicles σ_1,σ_2 and pedestrian stream σ_5, occurs during this signal. Although green light signal is for σ_5 , assuming that straight moving on red is allowed at the approach leads to conflicts.

3.4   Conflict Pattern and Rules

A conflict may follow one of three patterns:

(1) A vehicle occupies the pedestrian's target cell
(2) Pedestrians are within the vehicle's target area
(3) Both the pedestrian and the vehicle move towards the same target.

As a result of the conflict patterns, conflict rules are assigned as follows:

(1) When a vehicle occupies a pedestrian's target cell, the vehicle has priority to cross. In a similar manner, if the target cell of the vehicle is occupied by pedestrians, priority will be given to the pedestrians.
(2) If the pedestrian and vehicle share the same target cell, meaning their next time-step overlaps, the system will randomly give one of them the priority of passage with a probability of 0.5.

There are typically two different types of boundary conditions in the cellular automata model used to study traffic flow theory: periodic boundary conditions and open boundary conditions. It is best to analyse the CA model using the periodic boundary condition. The analysis is applicable when the density is constant and the vehicles are in a metastable state, and it requires that the start and end sections be connected. However, under the open boundary condition, when the vehicles move into and out of the road, it accords with a specific probability distribution, which is more realistic than the periodic boundary condition, so the open condition is used in this model.

4   Simulation of the Model

In this simulation, the concept of Cellular Automata (CA) serves as the foundational framework for modeling the intricate dynamics between pedestrians and vehicles at a signalized intersection. The urban space is metaphorically transformed into a regular grid R, where each distinct area symbolizes a cell within the CA grid structure. Within this gridded environment, the simulation encapsulates a comprehensive set of finite states A to describe the various conditions of both pedestrians and vehicles. Pedestrian states range from dynamic movements and waiting phases to conflicted situations, mirroring the finite states within the CA framework. Similarly, vehicles navigate through states encompassing motion, halting, and potential conflicts. The initiation of the simulation corresponds to the establishment of the initial configuration a_0, delineating the starting positions and states of pedestrians and vehicles. As the simulation unfolds over time, the concept of neighborhood M emerges organically, delineated by the proximity of pedestrians and vehicles. This proximity defines the set of potential conflicts and interactions, adhering to the CA principle of a neighborhood set.

The crux of the simulation lies in the manifestation of a transition function f, where a set of predefined rules govern the evolution of states for both pedestrians and vehicles. For instance, the transition from a waiting to a moving state for pedestrians is contingent on the signal turning green. This rule-based transition function aligns with the CA paradigm, encapsulating the dynamic and rule-driven nature inherent in cellular automata. Thus, by drawing parallels between the CA definition and the simulation results, we underscore how the structured and rule-based nature of Cellular Automata provides a robust framework for accurately modeling the intricate dynamics and interactions between pedestrians and vehicles in an urban setting. The simulation not only captures the essence of a regular grid with finite states but also highlights how these elements evolve over time through a defined transition function, reinforcing the utility and applicability of the CA framework in analyzing complex urban traffic scenarios.

This model is based on two assumptions:

(1) Traffic rules should be followed by both pedestrians and vehicles without engaging in intentional red light running. Only pedestrians who are still crossing at the end of the green light exhibit the red light running behaviour.
(2) Since pedestrian conflicts cannot affect safety, this model neglects them.

The model uses Python as a simulation platform. The flowchart in Fig. 5 depicts the process for simulating conflicts. Among them, the *vehicle decision* 1 denotes whether pedestrians are present in the target area, while the *vehicle decision* 2 denotes whether pedestrians are approaching the target area. The logic behind the pedestrians in *pedestrian decision* 1 and *pedestrian decision* 2 is the same (Fig. 6).

4.1   Simulation Results

Algorithm 1. Relationship Analysis of Pedestrian Flow, Vehicle Flow, and Conflicts

---

```
 1: procedure SIMULATECONFLICTS(pedestrian_flow, vehicle_flow)
 2:     conflict_events ← 0
 3:     Simulate conflicts based on model
 4:     for simulation_round in range(num_simulations) do
 5:         conflict_events += SimulateConflictEvent(pedestrian_flow, vehicle_flow)
 6:     end for
 7:     return conflict_events/num_simulations
 8: end procedure
 9:
10: procedure GENERATESIMULATIONDATA
11:     Generate pedestrian_flows and vehicle_flows data
12:     Create grid of flow combinations
13:     for each combination (ped_flow, veh_flow) in flow_combinations do
14:         conflict_count ← SimulateConflicts(ped_flow, veh_flow)
15:         Plot 3D point (ped_flow, veh_flow, conflict_count)
16:     end for
17: end procedure
```

The above Python code visualizes the relationship among pedestrian flow, vehicle flow, and the number of conflict events at a signalized intersection using a 3D scatter plot. The code defines a simulation function, simulate conflicts, that

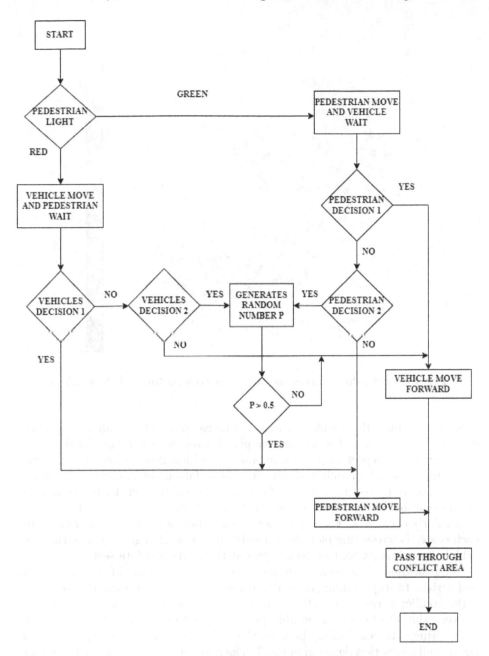

Fig. 6. The flowchart illustrating the simulation process

takes pedestrian flow and vehicle flow as input and returns the number of conflict events based on a simulated conflict model. Sample data is generated for pedestrian flows and vehicle flows, creating a grid of flow combinations. The

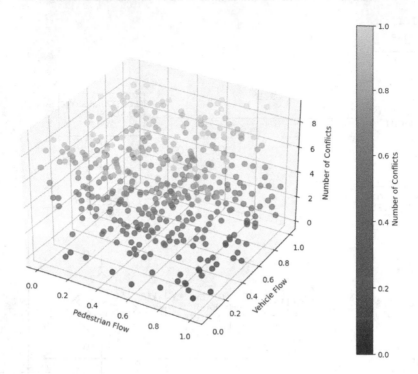

Fig. 7. Relationship between the pedestrian/vehicle flow and the conflicts

code then applies the simulation function to calculate the number of conflicts for each combination. The 3D scatter plot is created using Matplotlib, where the x and y axes represent pedestrian flow and vehicle flow, respectively, and the z-axis represents the number of conflicts. Each data point in the scatter plot is colored according to the intensity of conflicts, providing a visual representation of the relationship between pedestrian flow, vehicle flow, and conflict events. The colorbar serves as a reference for the number of conflicts associated with each color. The resulting plot offers insights into how changes in pedestrian and vehicle flows impact conflict occurrences at the signalized intersection.

The duration of the pedestrian conflict delay is the amount of time it takes for pedestrians to stop avoiding oncoming traffic. Pedestrians are free to cross again if the conflict is resolved. Actually, if the delay was excessive and the distance between the vehicles was acceptable, people on foot would cross the street without waiting. The relationship between the pedestrian flow, the vehicle flow, and the overall conflicts is depicted in Fig. 7. The conflict delay goes up as the vehicle density goes up for a certain pedestrian density. When the vehicle density is the same, the delay rises until the pedestrian arrival rate equals 0.04. Additionally, this pattern is consistent with the number of pedestrians who are involved in accidents with moving vehicles. When the pedestrian density is insufficient, more conflicts result from an increase in pedestrian flow. As pedestrians move and as

Algorithm 2. Relationship between the pedestrian/vehicle flow and the total conflict delay

---

1: **Inputs:** Pedestrian density, Vehicle density, Pedestrian arrival rate
2: **function** SIMULATECONFLICTDELAY(pedestrian_density, vehicle_density, pedestrian_arrival_rate)
3:     # Insert your simulation logic here
4:     # This could involve modeling the delay based on the provided details
5:     # For illustration purposes, a simple calculation is done
6:     $conflict\_delay$ ← pedestrian_density × vehicle_density × pedestrian_arrival_rate
7:     **return** $conflict\_delay$
8: **end function**
9: PedestrianDensities ← linspace(0, 1, 100)
10: VehicleDensities ← linspace(0, 1, 100)
11: PedestrianArrivalRate ← 0.04
12: Data ← []
13: **for** PedestrianDensity **in** PedestrianDensities **do**
14:     **for** VehicleDensity **in** VehicleDensities **do**
15:         ConflictDelay ← SIMULATECONFLICTDELAY(PedestrianDensity, VehicleDensity, PedestrianArrivalRate)
16:         Data.append((PedestrianDensity, VehicleDensity, ConflictDelay))
17:     **end for**
18: **end for**
19: x, y, z ← UNZIP(Data)
20: # Create 3D scatter plot with colorful points based on conflict delay
21: # (Insert appropriate LaTeX code for plotting)
22: # Set labels and title
23: # (Insert appropriate LaTeX code for labeling)
24: # Add colorbar
25: # (Insert appropriate LaTeX code for colorbar)

---

density rises even further, the likelihood that they would be stopped by moving vehicles while crossing decreases, and a downward trend is evident. But as the flow of pedestrians increases consistently, the quantity of pedestrians becomes the key element. Figure 8 shows the Relationship between the pedestrian/vehicle flow and the total conflict delay. As more and more people fail to cross at a green light and proceed through a red light, conflicts with passing vehicles increase. According to intuition, as each pedestrian's average delay rises and the number of conflicts rises, the overall delay also does. Meanwhile, this pattern is consistent with the proportion of pedestrians involved in collisions with moving vehicles.

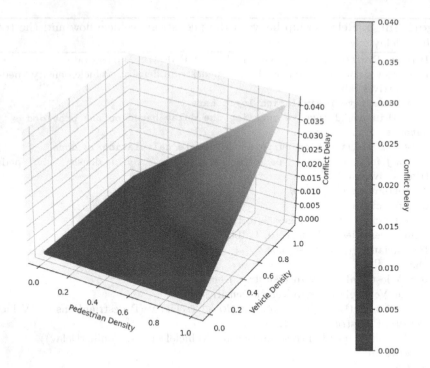

Fig. 8. Relationship between the pedestrian/vehicle flow and the total conflict delay

5 Conclusion

In this paper, we introduced a novel technique for examining collisions between pedestrians and moving vehicles at signalized intersections. Our approach is based on a detailed cellular automaton model, which allowed us to elaborate on pedestrian conflicts and associated delays. The simulation results revealed key insights into the dynamics of traffic flow, highlighting the impact of vehicle density on delay while showcasing fluctuations with varying pedestrian densities.

5.1 Key Findings

1. **Delay and Density Relationship:** The simulation demonstrated a direct relationship between delay and vehicle density, indicating that as the density of vehicles increases, so does the delay. This insight is crucial for understanding the potential challenges in traffic efficiency at signalized intersections.
2. **Pedestrian Density Fluctuations:** Interestingly, the delay tended to fluctuate with changes in pedestrian density. This observation underscores the complex interplay between pedestrian and vehicle flows, suggesting that factors beyond vehicle density also influence delay.
3. **Conflict Events Defined:** We defined and investigated conflict events, shedding light on specific scenarios where collisions between pedestrians and vehi-

cles are likely to occur. This understanding lays the groundwork for targeted interventions to enhance safety at signalized intersections.

4. **Useful Foundation for Research:** The proposed cellular automaton model serves as a valuable foundation for future research endeavors. It provides a robust framework for investigating how different pedestrian crossing designs impact both traffic efficiency and safety.

5.2 Future Work

1. **Pedestrian Crossing Designs:** Explore how various pedestrian crossing designs, such as signal timing adjustments or infrastructure modifications, can influence traffic efficiency and safety.
2. **Advanced Conflict Mitigation Strategies:** Investigate and develop advanced conflict mitigation strategies to further reduce delays and enhance the overall safety of mixed traffic flow.
3. **Real-world Validation:** Validate the findings of the cellular automaton model by comparing simulation results with real-world data from signalized intersections. This step would enhance the reliability and applicability of the model.
4. **Dynamic Signal Control Algorithms** Implement and assess the effectiveness of dynamic signal control algorithms that adapt in real-time based on varying pedestrian and vehicle densities.
5. **Integration with Traffic Management Systems:** Explore the integration of the proposed model with existing or future traffic management systems to develop practical and effective solutions for urban traffic challenges.

Overall, our study contributes valuable insights into the complex interactions between pedestrians and vehicles at signalized intersections. The defined conflict events and observed density-dependent delay trends pave the way for future research and interventions aimed at optimizing both traffic efficiency and safety in urban areas.

References

1. Blue, V.J., Adler, J.L.: Cellular automata microsimulation for modeling bi-directional pedestrian walkways. Transp. Res. B **35**(3), 293–312 (2001)
2. Brockfeld, E., Barlovic, R., Schadschneider, A., Schreckenberg, M.: Optimizing traffic lights in a cellular automaton model for city traffic. Phys. Rev. E. **64**(5), 056132 (2001)
3. Burks, A. W.: Essays on Cellular Automata. University of Illinois Press, Champaign (1970)
4. Chai, W., Wong, S.C.: Micro-simulation of vehicle conflicts involving right-turn vehicles at signalized intersections based on cellular automata. Transp. Res. Part B: Methodol. **67**, 202–222 (2014)
5. Garder, P.: Pedestrian safety at traffic signals - a study carried out with the help of a traffic conflicts technique. Accid. Anal. Prev. **21**(5), 435–444 (1989)

6. Gipps, P.G., Marksjo, B.: A micro-simulation model for pedestrian flows. Math. Comput. Simul. **27**(2), 95–105 (1985)

7. Jiang, Z., Wang, W., Zhu, M., Zhang, Y., Wu, J.: Modeling Traffic Conflicts between Pedestrians and Vehicles at Signalized Intersection based on the Cellular Automata. Transp. Res. Part D: Transp. Environ. **81**, 322–336 (2019)

8. Maerivoet, S., DeMoor, B.: Cellular automata models of road traffic. Phys. Rep. **419**(1), 1–64 (2005)

9. Nagel, K., Schreckenberg, M.: A cellular automaton model for freeway traffic. J. Phys. **2**(12), 2221–2228 (1992)

10. Narendra, K.A., Thathachar, M.A.L.: Learning Automata: An Introduction. Prentice Hall, New Jersey (1989)

11. Rahman, M.H., Rahman, M.M., Uddin, M.S., Mahmud, M.S.: Modeling and performance evaluation of a signalized intersection using cellular automata. Transport. Res. F: Traffic Psychol. Behav. **84**, 25–44 (2020)

12. Rajeswaran, S., Rajasekaran, S.: A study of vehicular traffic flow modeling based on modified cellular automata. IOSR J. Math. **4**, 32–38 (2013)

13. Von Neumann, V.: Theory of Self-Reproducing Automata. University of Illinois Press, (edited and completed by A. W.Burks) (1966)

14. Li, X.M., Yan, X.D., Wang, J.F.: Modeling traffic conflicts between pedestrians and vehicles at signalized intersection based on the cellular automata. Appl. Mech. Mater. **178**, 1881–1886 (2012)

15. Yang Yue, Sida Luo, Tiamming Luo.: Micro-simulation model of two-lane freeway vehicle for obtaining traffic flow characteristics including safety condition. J. Mod. Transp. **24**(3), 187–195 (2016)

16. Zhou, W., Zhu, M., Wu, J., Wang, W.: Using cellular automata to investigate pedestrian conflicts with vehicles in crosswalk at signalized intersection. Discrete Dyn. Nat. Soc. 963172 (2012)

Fire Spread Modeling Using Probabilistic Cellular Automata

Rohit Ghosh[1]([✉]), Jishnu Adhikary[1], and Rezki Chemlal[2]

[1] Government College of Engineering and Ceramic Technology, 73, Abinash Chandra Banerjee Lane, Kolkata 700 010, India
rohitghosh76@gmail.com
[2] Laboratory of Applied Mathematics, University of Béjaïa, Béjaïa, Algeria

Abstract. A cellular automaton (CA)-based modeling approach to simulate wildfire spread, emphasizing its strengths in capturing complex fire dynamics and its integration with geographic information systems (GIS). The model introduces an enhanced CA-based methodology for wildfire prediction, emphasizing interactions between neighboring cells and incorporating major determinants of fire spread, including wind direction, wind speed, and vegetation density, while also accounting for spotting and probabilistic transitions between states in the model to mirror real world fire behavior. This methodology is applied to case studies of the 1990 wildfire on Spetses Island, Greece, offering insights into the effects of terrain on fire spread, as well as the 2021 Evia Island wildfire in Greece, demonstrating the model's accuracy in simulating real-world wildfire scenarios.

Keywords: Cellular automata (CA) · Probabilistic Cellular Automata · Forest fires · Fire spread Modeling

1 Introduction

Forest fires have long posed significant threats, causing extensive damage to ecosystems, communities, and leading to far-reaching ecological and socio-economic consequences [1]. An alarming illustration of this destructive potential unfolded in August 2007 in Peloponnese, Greece when the nation faced its most severe forest fire catastrophe in a century, devastating approximately 2000 square kilometers of forest and agricultural land [1]. The urgency to develop and implement effective strategies to combat forest wildfires is intensifying as their frequency rises. Strategies to counteract wildfires fall into two categories: preventive measures designed to reduce the likelihood of fire ignition and immediate operational interventions in the event of a wildfire outbreak, involving the efficient allocation of defensive mechanisms and the swift evacuation of vulnerable communities [4].

Developing a predictive model for wildfire expansion necessitates consideration of external environmental factors like meteorological conditions and specific terrain

---

* This work is carried out as a project in the Summer School on Cellular Automata Technology 2023.

M. Dalui et al. (Eds.): ASCAT 2024, CCIS 2021, pp. 45–55, 2024.
https://doi.org/10.1007/978-3-031-56943-2_4

attributes [3]. Critical factors influencing the rate and pattern of a forest fire's progression encompass variables such as fuel type (vegetation classification), humidity, wind speed and direction, forest topography (slope and natural barriers), fuel density (vegetation thickness), and spotting—where burning materials are carried by the wind or other vectors, potentially extending the fire beyond its immediate front [2].

Constructing a comprehensive mathematical model for wildfire spread is a significant challenge, and it has received substantial attention in the literature, leading to various models. Notably, Rothermel's pioneering work [7, 8] stands out. Rothermel's research established dynamic equations characterizing the maximum fire spread rate through laboratory experiments. Rothermel's equations have subsequently formed the basis for various approaches, broadly categorized into two spatial representations. The first revolves around continuous planes, assuming the fire front navigates an uninterrupted landscape in an elliptical trajectory. However, solving the associated partial differential equations can be computationally demanding. The second approach, grid-based models, offers quicker computational solutions. An alternative yet efficient approach is presented [12, 13].

Within grid-based models, two subcategories emerge. The first uses the Bond–Percolation method, employing historical data for inter-grid fire spread probabilities. However, it falls short in capturing fire dynamics. The second subcategory centers on cellular automata (CA) methodology, offering a more dynamic approach with a discrete grid divided into cells governed by evolving variables and interactions with neighbouring cells [4]. This, combined with seamless integration with geographic information systems (GIS) and other data sources, makes CA an attractive candidate for modeling intricate wildfire behaviours [9, 10].

This study introduces an enhanced CA-based methodology for predicting wildfire spread, emphasizing interactions between neighbouring cells and incorporating major determinants of fire spread, including spotting. The model is applied to simulate a wildfire's progression in Spetses in 1990, which offers insights into the effects of terrain slope on fire spread. Additionally, the model's parameters are fine-tuned through a non-linear optimization approach, comparing the model's outputs with actual fire front data [4]. This work underscores the potential of CA-based models in effectively predicting wildfire dynamics and informing wildfire management strategies, especially in light of recent global fire incidents.

2 CA-Based Methodology for Wildfire Prediction

The predictive methodology for wildfire spread presented here employs a two-dimensional grid that subdivides the forest area into a multitude of cells. Each cell represents a small land patch, chosen to be square in shape, providing eight possible directions for fire propagation (refer to Fig. 2). Certain researchers have used [11] pentagonal or hexagonal cells, as they provide a more precise representation of spatial fire behaviour. However, it's important to acknowledge that such a choice significantly escalates computational complexity. In this paper, we favour the use of a square grid for its ability to simplify calculations.

2.1 Grid Definition

In the simulation, each cell on the 2D grid is associated with a specific state, and these states are represented by various categories. Here are the details of each state and its corresponding category:

- TREE: Contains unignited forest fuel.
- BURNING: Actively burning forest fuel.
- BURNING DURATION: Once the tree cell has been ignited by the fire, the cell transitions to this category, and the duration can be controlled probabilistically.
- EMPTY: Urban and non-vegetated areas with no forest fuel.
- WATER: Non-combustible water bodies.
- CITY: Urban zones or settlements within the simulation. These regions often have non-combustible properties and serve as obstacles to the fire's propagation.

The state of each cell is expressed as an element in a matrix denoted as the state matrix, specifically utilizing the Moore matrix to represent the directions of fire spread. An illustration of an area comprising 25 cells encoded in matrix form (Fig. 1).

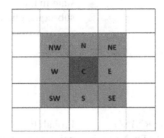

Fig. 1. Possible directions of fire propagation on a square grid.

2.2 Rules of Evolution

The CA model employs discrete states to characterize the evolution of each cell (i, j) at a given time step t. The Table 1 outlines specific rules that determine how cells behave and transition within the model, significantly impacting the dynamics of fire spread.

The local rule f can be defined as:

$$f : S^t_{(i,j)} \times S^t_{N(i,j)} \rightarrow S^{t+1}_{(i,j)}$$

- f is a function that takes two arguments:
- $S^t_{(i,j)}$: The present state of the central cell at position (i, j) at time t.
- $S^t_{N(i,j)}$: The states of the neighbouring cells in the neighbourhood of cell (i, j) at time t.

The function f determines the future state $S^{t+1}_{(i,j)}$ of the central cell at position (i, j) at time t + 1 based on its present state and the states of its neighbouring cells [14].

Table 1. CA Fire Spread Model Transition Rules

Rule	Current State	Next State	Description	Example
Rule 1	1 (TREE)	1 (TREE)	Areas that contain unignited vegetation or are devoid of burning fuel remain unchanged	If a cell contains unignited vegetation, it remains in the same state in the next time step
Rule 2	1 (TREE)	2 (BURNING)	Probability of tree ignition determined by a sigmoid function considering burning neighbors, vegetation density, and other factors	If burning neighbors exceed a threshold, 'TREE' transitions. Probability of 'TREE' catching fire can be dynamically adjusted
Rule 3	2 (BURNING)	3 (BURNING DURATION)	Actively burning cells transition to the "burning duration" state in the subsequent time step	If a cell is actively burning in the current time step, it turns into the "burning duration" state on the likelihood of neighboring cells (Moore neighborhood) in the next time step
Rule 4	3 (BURNING DURATION)	3 (BURNING DURATION) Varies based on pb	Probability of neighbouring cells transitioning to the "burning" state	If a cell is actively burning, there's a probability (pb) that one or more of its neighbouring cells (i ± 1, j ± 1) will also be on fire in the next time step
Rule 5	3 (BURNING DURATION)	(EMPTY)	Cells that have experienced complete combustion remain unchanged	If a cell has already burned, it remains in the same state in the next time step

2.3   Factors Influencing Fire Spread

Our methodology comprehensively considers variables that profoundly impact the shape and rate of wildfire propagation:

1. **Type and Density of Vegetation:** Vegetation's type (e.g., agricultural, shrubs, pine trees) and density (sparse to dense) are very important. We classify these into discrete categories, assigning probabilities (p_{veg} and p_{den}) that influence fire spread. We also

used a mix of empty cells and tree cells to control vegetation density, thereby altering fuel load and fire propagation potential.

2. **Wind Speed and Direction:** Wind's role in fire dynamics is critical. We utilize an adaptable empirical equation to calculate fire spread probability (p_w) by incorporating wind velocity and direction. This approach accommodates various wind directions, enhancing model realism.

3. **Spotting Effect:** Fire spotting, where embers ignite new fires ahead of the main fire, is incorporated indirectly through the combined influence of wind and ignition sources.

3 CA-Based Simulation

The proposed model harnesses the flexibility of probabilistic Cellular Automaton (CA) techniques to dynamically simulate the intricate dynamics of forest fires across diverse landscapes. In this paradigm, the landscape is effectively partitioned into cells, forming a matrix where each cell possesses distinct attributes. This 2D grid serves as the spatial foundation upon which the probabilistic Cellular Automaton (CA) operates.

The essence of the probabilistic CA model lies in its ability to incorporate stochastic elements that mimic real-world fire dynamics. Each cell within the landscape possesses an inherent state, such as 'EMPTY', 'TREE', 'BURNING', 'BURNING DURATION', 'WATER', or 'CITY'. The transitions between these states are governed by a set of probabilistic rules, driven by factors such as the presence of neighbouring burning cells, the likelihood of lightning strikes $p_{lightning}$ and the impact of wind effects $Wind_{Speed}$ and $Wind_{Direction}$.

This probabilistic approach introduces a layer of uncertainty into the simulation, mirroring the unpredictable nature of fire behaviour in actual forest environments. The transition probabilities are calculated based on the state of neighbouring cells and external influencing factors. For instance, if the sum of burning neighbouring cells exceeds a predefined threshold (threshold), a 'TREE' cell is ignited, transitioning to the 'BURN-ING' state. These probabilities can be dynamically adjusted to reflect changes in factors such as fuel density, moisture content, wind speed, topographical features, and even localized climate patterns.

The probabilistic rules' variability is a foundational strength of the model. For instance, the threshold required for ignition of a 'TREE' cell can be dynamically modified to account for varying levels of available fuel, fostering accurate depictions of real-world fire susceptibility. Likewise, the likelihood of lightning strikes ($p_{lightning}$) can be tuned to accommodate differing weather conditions and historical lightning activity in the region.

Incorporating wind effects is another illustrative example of adaptability. The wind's influence on fire spread, quantified by $Wind_{Speed}$ and $Wind_{Direction}$ can be fine-tuned to mirror localized wind patterns, resulting in more precise predictions of fire behaviour. This approach enables the model to capture the intricate interplay between variables, further enhancing its predictive capabilities.

Moreover, the transition probabilities between states are amenable to customization. The model can be configured to account for different burning durations, with probabilities adjusted to represent a variety of fire intensities. This flexibility extends to cell states such as 'BURNING DURATION', where the transition probability can mirror the actual likelihood of a fire sustaining itself over time.

This probabilistic CA model embraces the randomness of fire-related events, capturing the interplay of neighbouring cell states, probabilistic influences, and the evolving nature of fire spread. It works as an excellent tool for simulating complex wildfire scenarios and enhancing our understanding of fire behaviour within diverse landscapes. In essence, the spatial distribution of probabilities, encoded within the network topology, facilitates the realistic emulation of fire dynamics, advancing the predictive capabilities of fire management strategies.

3.1 Tree Ignition Probability

Trees have a probability of catching fire, which is determined by a sigmoid function. This function maps the number of burning neighbours to a probability value between 0 and 1. As the number of burning neighbours increases, the probability of a tree catching fire also increases. The *sigmoid function* is defined as:

$$P_{ignition}(burning_{neighbours}) = {}^1\!/\!(1 + exp(-k * (burning_{neighbours} - threshold)))$$

where:

- $P_{ignition}$ is the probability of a cell catching fire.
- $burning_{neighbours}$ is the number of burning neighbors of the cell.
- $threshold$ is a parameter that determines when ignition is likely.
- k is a scaling parameter that controls the steepness of the sigmoid curve.

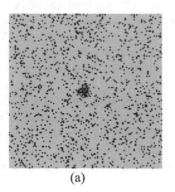
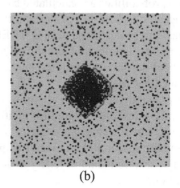

(a)   (b)

Fig. 2. (a) Initial configuration of the CA when it is ignited; (b) Configuration shows the spread of fire with 90% Vegetation Density

3.2 Wind Effects

The simulator takes into account the influence of wind on fire spread. Wind direction and speed affect the probability of fire spreading in a particular direction. The wind effects can be incorporated using a differential equation for fire spread (Fig. 3):

$$dF/dt = k * Wind_{Speed} * Wind_{Direction} - m * F$$

where:

- F represents the fraction of cells that are on fire in the neighbourhood of the current cell.
- dF/dt is the rate of change of the fraction of cells on fire.
- $Wind_{Speed}$ is the magnitude of the wind.
- $Wind_{Direction}$ is a vector representing the wind direction.
- k is a constant that determines the influence of wind on fire spread.
- m is a constant representing the fire extinguishing rate.

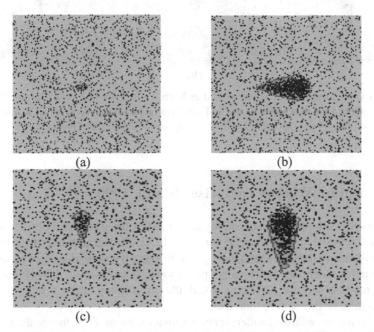

Fig. 3. (a) Initial CA Configuration with Eastward Wind Direction; (b) Fire Spread Progression Eastwards; (c) Initial CA Configuration with Southward Wind Direction; (d) Fire Spread Progression Southwards

3.3   Calculation of Fire Spread Probability (p_b)

The probability of fire propagation pb is synthesized considering the above variables. We factor in a constant probability po that a cell adjacent to a burning cell catches fire. This adjusted probability takes into account the combined effect of vegetation, density, wind, and terrain:

$$p_b = p_o(1 + p_{veg})(1 + p_{den})p_w$$

where:

po represents the baseline probability of fire propagation, derived from empirical data in the absence of wind effects, on flat terrain, and for a specific combination of vegetation density and type. This foundational probability serves as a reference point for further calculations. The components p_{den}, p_{veg} and p_w correspond respectively to the influences of vegetation density, vegetation type, and wind conditions (speed and direction) on the fire spread dynamics (Fig. 4).

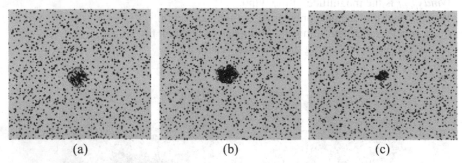

(a)   (b)   (c)

Fig. 4. (a) Configuration shows Fire Spread at Each Time Step with 90% Burning Probability; (b) Configuration shows Fire Spread at Each Time Step with 50% Burning Probability; (c) Configuration Shows Fire Spread at Each Time Step with 30% Burning Probability

4   Case Study: Wildfire on Spetses Island, 1990

Our proposed methodology was employed to predict the progression of a significant wildfire that ravaged Spetses Island on August 1, 1990. The fire originated from unknown sources within the island's forest, igniting near its central region. Propelled by moderate to strong north winds, the fire rapidly spread towards the south. Approximately 11 h later, the fire was successfully extinguished after consuming nearly 6 km2, equivalent to about a third of the island's total area.

We selected this wildfire incident for two primary reasons. First, the incident's details, such as the extent of burnt area and time to extinguishment, were meticulously documented by local authorities. Second, the island's distinct topographical features, provided an ideal testing ground for evaluating the efficacy of our proposed Cellular Automaton (CA)-based model.

Initiating the application of our methodology involved generating matrices reflecting vegetation density and type. To achieve this, we created shape-files coding the island's vegetation type and density, drawing from pre-wildfire photomaps. Additionally, a shape-file delineating the burned area was generated. These digital datasets were then superimposed onto the grid to construct a vector data file. We adopted a square grid, to ensure precise representation of geographical data. From this vector data file, we derived matrices corresponding to vegetation density and type, alongside a vegetation density matrix and the aggregate burnt area.

Spetses Island hosts three vegetation types: Aleppo-pine trees, shrubs, and agricultural regions. The vegetation density was classified into three distinct categories: sparse, normal, and dense. Drawing from a combination of well-documented incident details and modern GIS tools, this case study offers a tangible platform to evaluate the capabilities of our CA-based model, as applied to real-world wildfire scenarios (Fig. 5).

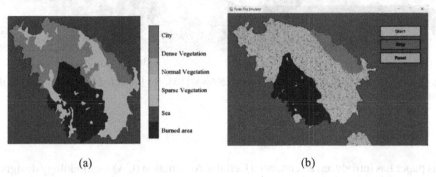

(a)   (b)

Fig. 5. (a) Actual Burned Area (Recreated from Photomap) [5]; (b) Predicted Burned Area from the Simulation.

5   Case Study of the 2021 Evia Island Wildfire, Greece

For our analysis we took as paradigm the case of the 2021 Greece wildfires, an illuminating case study emerged on Evia Island, Greece. Amidst a historic heatwave driving temperatures to an unprecedented 47.1 °C (116.8 °F), multiple wildfires ravaged the nation, inflicting immense damage. Against this backdrop, the largest wildfires tore through regions like Attica, Olympia, Messenia, and most devastatingly, northern Euboea. The second one suffered the most, with more than 50,000 hectares burned. Despite these difficulties, our method using Cellular Automaton (CA) was highly accurate for predicting wildfires.

The fires of 2021 scarred approximately 125,000 hectares of forest and arable land, echoing the 2007 Greek forest fires in magnitude. In this turbulent weather, a separate blaze struck Rhodes Island, prompting mass evacuations and disruptions. Amid this destruction, the 2021 Evia Island wildfire case study showed how effective our CA-based model can be.

By carefully combining data about incidents, modern GIS tools [9, 10] and the island's detailed terrain, our model demonstrated it can accurately simulate real wildfires. As climate change makes wildfires more dangerous, CA-based methods show us a way to predict and manage wildfires ahead of time, helping us adapt to changing challenges (Fig. 6).

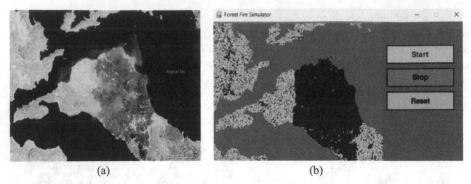

(a)   (b)

Fig. 6. (a) Actual Satellite Image Captured by NASA [15]; (b) Predicted Burned Area from the Simulation.

6 Conclusion

This paper has introduced an enhanced Cellular Automaton (CA) methodology designed to dynamically forecast wildfire spread. This approach factors in a multitude of influences on fire propagation, incorporating both meteorological and spatial geographical variables.

The practical application of our methodology entailed the simulation of a genuine wildfire that engulfed a renowned Greek island, destroying nearly half of its forested land. The island's intricate terrain, featuring abrupt elevation changes and diverse vegetation characteristics, provided an ideal testing ground. To optimize the model's performance, a non-linear optimization technique was employed to fine-tune certain coefficients, minimizing the disparity between actual and predicted burnt areas. This optimization process involves running the simulation multiple times with different sets of parameter values, comparing the simulated results to real data, and iteratively adjusting the parameters to minimize the difference between the two. Encouragingly, the simulation outcomes closely aligned with real-world observations, affirming the potential of our methodology in effectively predicting fire spread across heterogeneous landscapes.

While the model has demonstrated promising results, its robustness will undergo further validation through application to diverse large-scale, real-world fire incidents. Ongoing efforts are directed towards improving the model's accuracy by incorporating various factors such as different types of vegetation, altitude, elevation, and humidity. Also recognizing the significant role of near-surface airflow patterns in fire front evolution, further improvements in the model is anticipated to enhance prediction accuracy.

In light of the accomplished work, future works should integrate additional parameters such as humidity, terrain features, and vegetation density. By expanding the model's scope, we aim to achieve higher simulation accuracy and extend its applicability to a broader range of wildfire scenarios. This approach holds the promise of improving wildfire prediction and management strategies, offering proactive solutions to mitigate the devastation caused by such events.

References

1. Pyne, S.J., Andrews, P.L., Laven, R.D.: Introduction to Wildland Fire. New York, USA, Wiley (1996)
2. Albini, F.A., Brown, J.K.: Mathematical modelling and predicting wildland fire effects. Combust. Explosion Shock Waves **32**, 520–533 (1996)
3. Alexandridis, A., Russo, L., Vakalis, D., Siettos, C.I.: Simulation of wildland fires in large-scale heterogeneous environments. Chem. Eng. Trans. **24**, 433–438 (2011)
4. Alexandridis, A., Russo, L., Vakalis, D., Bafas, G.V., Siettos, C.I.: Wildland fire spread modelling using cellular automata: evolution in large-scale spatially heterogeneous environments under fire suppression tactics. Int. J. Wildland Fire **20**, 633–647 (2011)
5. Alexandridis, A., Vakalis, D., Siettos, C.I., Bafas, G.V.: A cellular automata model for forest fire spread prediction: the case of the wildfire that swept through Spetses Island in 1990. Appl. Math. Comput. **204**, 191–201 (2008)
6. Berjak, S.G., Hearne, J.H.: An improved cellular automaton model for simulating fire in a spatially heterogeneous Savanna system. Ecol. Modell. **148**, 133–151 (2002)
7. Rothermel, R.: A mathematical Model for predicting fire spread in wildland fuels, Res. Pap. INT-115, Ogden, UT, US, Dept. of Agric., Forest Service, Intermountain Forest and Range Experiment Station (1972)
8. Rothermel, R.: How to predict the spread and intensity of forest fire and range fires, Gen. Tech. Rep. INT-143, Ogden, UT, US, Dept. of Agric., Forest Service, Intermountain Forest and Range Experiment Station (1983)
9. Vakalis, D., Sarimveis, H., Kiranoudis, C., Alexandridis, A., Bafas, G.: A GIS based operational system for wildland fire crisis management I. Mathematical modelling and simulation. Appl. Math. Model. **28**(4), 389–410 (2004). https://doi.org/10.1016/j.apm.2003.10.005
10. Yassemi, S., Dragicevic, S., Schmidt, M.: Design and implementation of an integrated GIS-based cellular automata model to characterize forest fire. Ecol. Modell. **210**, 71–84 (2008)
11. Yongzhong, Z., Feng, Z.D., Tao, H., Liyu, W., Kegong, L., Xin, D.: Simulating wildfire spreading processes in spatially heterogeneous landscapes using an improved cellular automaton model. In: IGARSS 2004, Proceedings of the 2004, pp. 3371–3374. IEEE International, (2004)
12. Karafyllidis, A.T.: A model for predicting forest fire spreading using cellular automata. Ecol. Modell. **99**(1), 87–97 (1997)
13. Li, X., Magill, W.: Modeling fire spread under environmental influence using a cellular automaton approach. Complex. Int. **8**, 1–14 (2001)
14. Almeida, R.M., Macau, E.E.N.: Stochastic cellular automata model for wildland fire spread dynamics. J. Phys. Conf. Ser. **285**, 012038 (2011)
15. NASA Earth Observatory: Fire Consumes Large Swaths of Greece. https://earthobservatory.nasa.gov/images/148682/fire-consumes-large-swaths-of-greece

Temporally Non-Uniform Cellular Automata as Pseudo-random Number Generator

Subrata Paul[1]([✉])[iD] and Kamalika Bhattacharjee[2][iD]

[1] Department of Information Technology, Indian Institute of Engineering Science and Technology, Shibpur, Howrah 711103, West Bengal, India
psubrata.it@gmail.com
[2] Department of Computer Science and Engineering, National Institute of Technology, Tiruchirappalli 620015, Tamil Nadu, India
kamalika.it@gmail.com

Abstract. In this work, we use Temporally Non-Uniform Cellular Automaton (t-NUCA) where, two rules are used temporally. That is, there exists a rule sequence, which specify the behavior of a t-NUCA. We explore the dynamics of t-NUCAs for different rule sequences and observe the dynamical behavior of each t-NUCA. In addition to that, the t-NUCAs which have no non-reachable configurations are detected. Among them, we identify a set of t-NUCAs which are surjective and chaotic in nature. We show that such t-NUCAs can be used to develop a pseudo-random number generator (PRNG).

Keywords: Temporally Non-Uniform Cellular Automata (t-NUCA) · Pseudo-Random Number Generator (PRNG) · Rule Sequence · Dieharder · NIST

1 Introduction

A string of bits is considered random in the classical sense if it cannot be identified with a string shorter than itself. In other words, it cannot be produced effectively by a program that is smaller than it [3]. Deterministic algorithms that produce a periodic sequence of random numbers are known as pseudo-random number generators (PRNGs). They extend a short random seed into a lengthy sequence of values that act in accordance with an 'actual' random sequence. For this, they are commonly known as pseudo-random numbers.

Cellular automata (CAs) have long been recognized as a good source of randomness and have several advantages over other methods, such as simplicity, local interactions, and parallelism, especially when used with hardware implementation. These advantages, together with pseudo-randomness and simplicity

This work is carried out as a project in the Summer School on Cellular Automata Technology 2023.

of scaling, have led to substantial study in CAs in fields such as cryptography, field programmable gate arrays, Monte Carlo simulations, VLSI circuit testing [2,9,15], etc.

Classical cellular automaton (CA) uses a single rule which is applied uniformly over the grid. This work uses temporally non-uniform cellular automata (t-NUCAs), where two rules are used temporally during evolution. There exists a rule sequence for a temporally non-uniform cellular automaton (t-NUCA), which indicates the order of the applied rules with respect to time. Based on the rule sequence, t-NUCAs are varied. Some variant of t-NUCAs are previously targeted to explore and use in different applications [10,11]. In this work, we target to use a variant of t-NUCAs to develop a pseudo-random number generator (PRNG).

In the following sections we explore the background of t-NUCAs and the features of a good pseudo-random number generator (PRNG). We map the required features with the t-NUCAs and identify some properties for finding some potential candidates. Then we develop some sample PRNGs based on t-NUCAs and show their performances.

2   Background

A discrete abstract model of dynamical systems known as cellular automaton (CA) is made up of a regular network of finite state machines. These machines are referred to as cells. A local rule updates a cell's subsequent states. The input of the local rule is considered to be the current states of a cell and its neighbors. The main objective of this study is to use cellular automata to develop a pseudo-random number generator (PRNG). Let us first discuss about pseudo-random number generator (PRNG) in the next section.

2.1   Pseudo-random Number Generator

A sequence of *random* numbers can be produced using a deterministic algorithm called a pseudo-random number generator. Formally, a PRNG can be defined in this way:

Definition 1. A pseudo-random number generator (PRNG) G is defined as $(\mathcal{S}, \mu, f, \mathcal{U}, g)$, where μ is the probability distribution on finite set of states $\mathcal{S}$ for the initial configuration, called seed. The transition function is denoted by $f : \mathcal{S} \to \mathcal{S}$. Here, $\mathcal{U}$ is the finite set of output states and the output function $g : \mathcal{S} \to \mathcal{U}$ [5].

A generator operates in this way:

1. The seed $s_0 \in \mathcal{S}$ is selected based on the probability distribution μ. Hence, $u_0 = g(s_0)$.
2. The states $s_i = f(s_{i-1})$ and the output states $u_i = g(s_i)$ for $i \in \mathbb{N}$. Here, the sequence is pseudo-random, and the outputs are pseudo-random numbers.

Hence, the parameters f, g and s_0 are used to classify PRNGs. As a result, it is assumed that the adversary is unaware of the technique when the randomness quality of a PRNG is identified.

In classical sense, a PRNG can generate uncorrelated and independent real numbers. The numbers are generally uniformly distributed, however, other probability distribution can also be applied [5]. The desirable characteristics that make those generated numbers random when developing a PRNG are uniformity, independence, large period, reproducibility, consistency, disjoint sub-sequence, portability, efficiency, coverage, spectral characteristics, and cryptographically secure [1]. Many of these features are connected. A decent PRNG's numbers should ideally meet each of these requirements. However, majority of the PRNGs do not have all of these characteristics, e.g., disjoint Sub-sequence mutations, cryptographically secure, etc. are frequently absent in the existing PRNGs. Still, many PRNGs are regarded as good by today's standards for utilization in the applications for which they are designed.

2.2   Performance Metrics

There are several statistical tests that seek to identify non-randomness using a variety of statistical tests. NIST and Dieharder are two such well-known test suites used in this work.

Dieharder Battery of Tests: This battery of tests is an extension of the original batttery of tests provided by George Marsaglia in 1996 [6] and serves as the fundamental testbed for PRNGs. A PRNG first takes a seed before generating numbers. In this case, these produced numbers are utilized to create a binary file that can be used to test the PRNG on Dieharder for a given seed. In this study, we have considered binary file sizes ranging from 1.2 GB to 12 GB, and one or more p-values are produced for each test. The test is deemed to have passed only when all its p-values fall between 0.025 and 0.975 [6]. There are 114 such tests in the testbed.

NIST Statistical Test Suite: A PRNG's cryptographic features are tested using the NIST test suite [12]. The significance threshold for this test is $\alpha = 0.01$. Let k be the minimum passing rate for a PRNG with a produced sample size of l, then the minimum number of sequences with p-values ≥ 0.01 has to exceed k to pass the test. In this study, we consider the seed length to be of 64 bits, and the input is a binary file with a size of 1.2 GB. There are 188 tests in the NIST implementation that we have utilized.

Space-Time Diagram: In Ref. [1], space-time diagrams have been introduced as a metric to measure performance of PRNGs. If the space-time diagram contains patterns, then the PRNG is not good. To draw a space-time diagram, the number of cells of the CA has to be fixed to some $n \in \mathbb{N}$. Here, the PRNG is tested on the space-time diagram by selecting a window of length w, initialized randomly, and setting the remaining cells $(n - w)$ to $0^{n-w-1}1$.

2.3 Temporally Non-uniform Cellular Automata

This work investigates the behavior of temporally non-uniform cellular automata (t-NUCAs), a special kind of cellular automata. A temporally non-uniform cellular automaton (t-NUCA) is a variant of non-uniform cellular automaton, where the CA uses more than one local rule to update its cells over time. Let us define the t-NUCA:

Definition 1. *A temporally non-uniform cellular automaton (t-NUCA) is a quadruple $(\mathcal{L}, \mathcal{N}, \mathcal{S}, \mathcal{R})$ [10] where,*

- $\mathcal{L} \subseteq \mathbb{Z}^D$, *where D is the dimension of the lattice $\mathcal{L}$, each element of $\mathcal{L}$ is called a cell.*
- $\mathcal{N} = (\boldsymbol{v}_1, \boldsymbol{v}_2, \cdots, \boldsymbol{v}_m)$ *is the neighborhood vectors. Let $\boldsymbol{v}$ is a cell of lattice $\mathcal{L}$ then $(\boldsymbol{v} + \boldsymbol{v}_i) \in \mathcal{L}$, where m denotes the number of neighbors.*
- $\mathcal{S}$ *is the set of states.*
- $\mathcal{R}$ *is a sequence of local rules, which are applied temporarily. That is, $\mathcal{R} = (\mathcal{R}_t)_{t \geq 1}$, where each $\mathcal{R}_t$ is a local rule and it is applied to all cells at time step t.*

In this work, we consider one-dimensional two-state cellular automata, where each rule follows three-neighborhood structures. That implies, $\mathcal{R}_t : \mathcal{S}^3 \to \mathcal{S}$ is applied to all cells at time step t, where $\mathcal{R} = (\mathcal{R}_t)_{t \geq 1}$ is a temporal rule sequence of local rules. If we consider only one rule in the entire rule sequence, this is similar to *Elementary Cellular Automata* (ECAs). When more than one rule is used in $\mathcal{R}$, the dynamics of the CA can be affected by the rule sequence $\mathcal{R}$. In this work, we consider two rules f and g, and a condition $\Theta(t)$. We consider, the proposed CAs are finite and follows periodic boundary condition, that is, the lattice $\mathcal{L} = \mathbb{Z}/n\mathbb{Z}$ where n is the number of cells. The rule sequence can be defined using $\Theta(t)$, that is, rule f is applied when $\Theta(t)$ holds, rule g is applied otherwise.

$$\mathcal{R}_t = \begin{cases} f & \Theta(t) \text{ holds} \\ g & \text{otherwise} \end{cases}$$

The t-NUCA can be identified using a notation $(f, g)[\mathcal{R}]$, where $\mathcal{R}$ is the rule sequence. It is difficult to mention whole sequence in this notation. This work considers an integer sequence, that denotes the time series when f is applied in the evolution. This sequence (integer series) can be associated with a lexicographical ordering number in OEIS database [16]. Instead of using the entire sequence to express the rules, we utilize the OEIS order number, for simplicity.

Example 1. Let us consider $\Theta(t)$ holds when $t \bmod 4 \neq 0$. Then the series is: $1, 2, 3, 5, 6, 7, \cdots$ (that is, OEIS $A042968$ [16]). That is, the rule f is applied when $t \bmod 4 \neq 0$. A t-NUCA that consists of this sequence and two rules (f and g) can be identified as $(f, g)[A042968]$.

2.4   T-NUCA Dynamics

The *t-NUCAs* can use different kind of rule sequences consisting of multiple rules. In this work, we consider only two rules (say f and g) in a rule sequence. We explore the space-time diagrams of different t-NUCAs to study the dynamics.

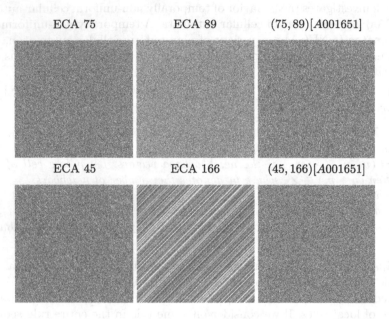

Fig. 1. Space-time diagram of two *t-NUCAs* $(75, 89)[A001651]$ and $(45, 166)[A001651]$. The CA considers *periodic* rule sequence, that is, *OEIS* $A001651$, where f is applied twice, and then the rule g is used once, and this repeats. The CA size is 500, and 500 iterations are shown. Here, the CA considers random initial configuration with the initial density as 0.5 (density of the cells with state 1).

In Fig. 1, we can observe that the dynamics of the *t-NUCA* using rules 75 and 89 is much different from the dynamics of the ECAs 75 and 89. One can see that, rule 89 is applied as an impurity on the dynamics of ECA 75. Hence the dynamics of 75 is perturbed. In Example 2, this abrupt changes in dynamics provides an insights that the impurity may increase (or decrease) the randomness of the system. Hence, this update scheme can be a ideal choice of developing a CA based pseudo-random number generator.

Example 2. Let us consider a *t-NUCA* with ECA rules 75 and 89. Assume, the sequence is: $75, 75, 89, 75, \cdots$ $(OEIS\ A001651)$. This implies, rule 75 is applied twice followed by a single 89, and this repeats. A snapshot of space-time diagram of this *t-NUCA* $(75, 89)[A001651]$ is shown in Fig. 1. Snapshot of space-time diagram of *t-NUCA* $(45, 166)[A001651]$ is also shown in Fig. 1.

As we use two rules, the dynamics for the individual rules influence the global behavior of the CA. The following theorems are studied in Ref. [10].

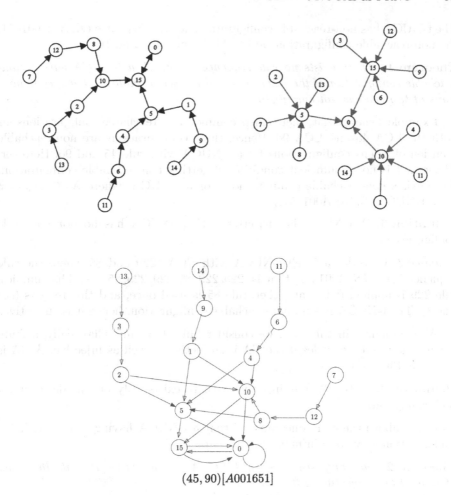

$(45, 90)[A001651]$

Fig. 2. Transition diagram of a 4-cell *t-NUCA* $(45, 90)[A001651]$. The rule sequence is *OEIS* $A001651$, where f is applied twice, and then the rule g is used once, and this repeats. Here, the transition is shown with black and blue arrows according to rules 45 and 90, respectively. The decimal representation of each configuration is shown in the corresponding node.

Definition 2. A configuration is said to be non-reachable if it is not reached during entire evolution of a t-NUCA.

Example 3. Let us assume a 4-cell t-NUCA consists of two ECAs 45 and 90. Intersection of the sets of non-reachable configuration of these two ECAs is non-empty. Now, assume that the t-NUCA follows a rule sequence of OEIS $A001651$, that implies, rule 45 is applied twice followed by a single 90, and this repeats. Figure 2 shows the transition diagram of this t-NUCA $((45, 90)[A001651])$, where black and blue arrow indicate the transitions by rule 45 and rule 90 respectively.

The t-NUCA has non-reachable configurations for the sequence $OEIS$ A001651: the non-reachable configurations are, $7, 11, 13$, and 14 (Fig. 2).

Theorem 1. *There exists no rule sequence for which a t-NUCA has no non-reachable configuration if the intersection of the sets of non-reachable configurations of two rules is not empty [10].*

Example 3 shows that, there exists common non-reachable configurations for both the ECA 45 and ECA 90. Hence, these configurations are non-reachable (Garden of Eden) configurations for a t-NUCA with rule 45 and 90. However, Theorem 1 is not a sufficient condition of getting non-reachable configuration; there exists non-reachable configurations for a t-NUCA, where $X_f \cap X_g = \emptyset$ (e.g., t-NUCA $(3, 15)[A001651]$).

Definition 3. A t-NUCA is surjective if the t-NUCA has no non-reachable configuration.

Example 4. Consider a 5-cell t-NUCA with ECAs 229 and 85 where the rule sequence is $OEIS$ A001651, that is, $229, 229, 85, 229, 229, 85, \cdots$. That implies, rule 229 is applied twice, and then rule 85 is used once, and this repeats (see Fig. 3). The t-NUCA has no non-reachable configuration, hence it is surjective.

As mentioned, in this wok, we consider finite CA only. Classically, a finite CA is able to create cycles if the CA is surjective as well as injective. A CA is injective if it is surjective.

Definition 4. A t-NUCA is injective if there exists only one predecessor for each configuration.

From the above theorem, one can find that a t-NUCA having a non-reachable configuration may contain many-to-one map.

Theorem 2. *For any sequence, a t-NUCA is surjective if both the rules (f and g) are surjective [10].*

Our target is to select the desired candidate ECAs such that the generated t-NUCA contains no non-reachable configuration. We consider the t-NUCAs where at least one rule is surjective. Then, we further investigate the dynamical properties of them.

3   t-NUCA as Pseudo-random Number Generator

Cellular automata are completely deterministic. The next configuration can be easily deduced from the previous configuration by applying rules. However, in a good PRNG based on CAs, the CAs should be autoplectic. In such kind of CAs, small changes in a local cell spread to the whole global system. There exists some properties that ensure this behavior of the system, these are, balancedness of the next state values in the rule, large cycle length in the configuration space, the rules being non-linear, a substantial flow of information in the CA, etc. These properties help to make a CA chaotic and a good candidate for PRNG. Hence, we identify the t-NUCAs, that hold the above properties.

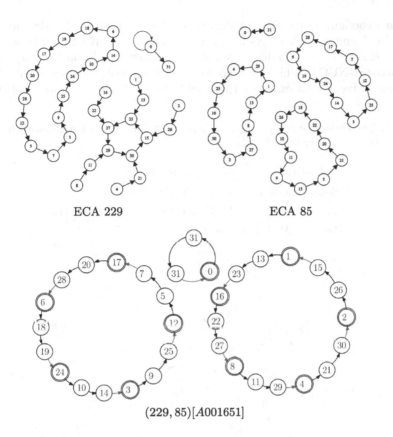

ECA 229   ECA 85

(229, 85)[A001651]

Fig. 3. Transition diagram of a 5-cell *t-NUCA* (229, 85)[A001651] [10]. (a) Transition diagram of ECA 229; (b) Transition diagram of ECA 85; (c) Transition diagram of the *t-NUCA* (229, 85)[A001651]

3.1   Finding the Desired Candidate T-NUCAs

To be a PRNG, first we need to make sure that the t-NUCA has no non-reachable configuration.

Theorem 3. *There exists a rule sequence for which a t-NUCA has no non-reachable configurations if the first rule in the rule sequence is surjective.*

Proof. Let us assume the rule f as the first rule used in a rule sequence and f is surjective. Let C is a set of all possible configuration and $C = G_f(C)$. Let us consider $(G_g^k \circ G_f^l)^+(C)$ *for* $k, l \geq 1$ is the general rule sequence. We consider the intermediate configuration that are reachable by f and g. As f is surjective, by f all possible configurations can be reached. This implies, the t-NUCA has no non-reachable configuration. Hence the proof.

For example, a t-NUCA consisting of ECAs 105 and 90, where rule sequence is *OEIS A001651* is a surjective t-NUCA (Fig. 4) as the t-NUCA contains no non-reachable configurations. For our desired t-NUCAs, we consider the surjective ECAs for f and all 256 ECAs for g. The list of surjective ECAs is shown in Table 1. However, this condition is not a necessary condition for surjectivity. There exists t-NUCAs which have no non-reachable configurations, even when f is not surjective; see for example, the 4-cell t-NUCA $(229, 85)[A001651]$ (Fig. 3).

Table 1. The surjective rules of ECAs. Here, all 256 ECA rules are considered, and the CA size n is greater than 3

CA size (n)	Surjective ECAs
Independent of n	15, 51, 85, 170, 204, 240
When n is odd	45, 75, 89, 101, 154, 166, 180, 210
When $n \neq 3k, k \in \mathbb{N}$	105, 150

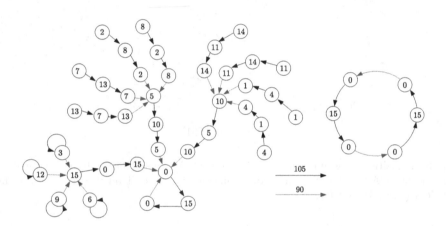

Fig. 4. Transition diagram of a 4-cell $t\text{-}NUCA$ $(105, 90)[A001651]$. Here, the first rule is a surjective rule. Black and blue arrows represent the transition by the rules 105 and 90 respectively

Theorem 4. *There exists finite t-NUCA where surjectivity does not imply injectivity.*

In Example 4, the t-NUCA contains no non-reachable configurations, however, it has many-to-one map for the configurations 0 and 31 (Fig. 3). This indicates for finite t-NUCA, surjectivity does not imply injectivity.

Another important property is, we consider the rules which are chaotic in nature. Hence, for rule f, we need to consider the ECAs which are surjective and chaotic (e.g. ECA 150). For rule g, we need to consider all the chaotic ECAs (for instance, ECAs $30, 45, 90, \cdots$) [7] to find the desired t-NUCAs.

We explore the possible candidate t-NUCAs based on the above proper-ties, and synthesize t-NUCAs which are good candidates for PRNGs. We iden-tify the t-NUCAs which are surjective and able to generate large cycles. We choose the *OEIS* sequences experimentally that can be used to develop a good PRNG. For example, let us assume a t-NUCA with two ECAs 30 and 150, and the rule sequence follows fibonacci sequence, that is, the series of sequence is $1, 2, 3, 5, 8, 13, 21, 34 \cdots$ (*OEIS* A000045). The t-NUCA ((30, 150)[A000045]) shows good performance as a pseudo-random number generator (Table 2).

Example 5. Let us consider a t-NUCA (150, 30)[A042968] that is the series of sequence is $1, 2, 3, 5, 6, 7, 9, \cdots$ (OEIS A042968) (see Fig. 6). This t-NUCA is surjective and chaotic in nature. Transition diagram of the 5-cell t-NUCA (150, 30)[A042968] is shown in Fig. 5. In this figure, we can observe that, all configurations (except 0 and 31) form a single connected graph and there exists no non-reachable configuration. This gives us an insight of getting large cycle using such a rule sequence. Space-time diagram of the t-NUCA is shown in Fig. 6. From Fig. 6.c, one can find that the dynamical behavior of the t-NUCA is unpredictable and chaotic in nature. Such t-NUCA has the potentiality to be used as a PRNG.

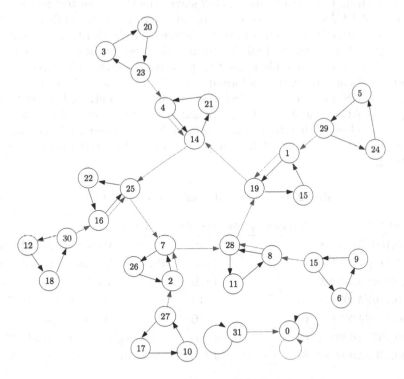

Fig. 5. Transition diagram of a *t-NUCA* (150, 30)[A042968] with lattice size 5. Config-urations are noted by their decimal equivalents with first bit as LSB

ECA 150   ECA 30 $(150, 30)[A042968]$

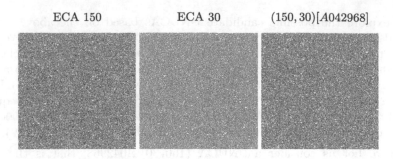

Fig. 6. Space-time diagram of a *t-NUCAs* $(150, 30)[A042968]$. The size of the CA is 500. Here, blue implies state 1 updated by ECA 150 and red implies state 1 updated by ECA 30 and white implies state 0

3.2   Performance Analysis

To check how good the proposed PRNGs are, we need to measure the randomness quality of the model over the performance metrics discussed in Sect. 2.2. Here, to reduce wastage of bits, we consider the CA size $n = 65$ only where the output is extracted from the first 64 bits. Table 2 shows the Dieharder test results of the candidate t-NUCAs, where the seed is considered as a 64 bit number. Similarly, Table 3 shows NIST test results of the candidate t-NUCAs for two rule sequences *OEIS* A005408 and *OEIS* A001651. Here also, a seed is considered as a 64 bit number. A 1.2 GB binary file is used to perform the tests. The t-NUCAs and the test results are shown in this format: the t-NUCA is identified as $(f, g)[$Rule sequence (OEIS number)$]$ and NIST test results are mentioned as: number of tests passed / total number of tests. Although the NIST test suite suffers from bugs [4], we observe that there exists some t-NUCA based PRNGs that show good performance and pass a lots of empirical tests. These test results are shown in Table 4.

Table 2. Dieharder test results of some t-NUCAs

t-NUCA	Number of bits	File size	Dieharder
$(15,180)[A005408]$	64	4 GB	Pass=55; Weak=15; Failed=44
$(15,210)[A005408]$	64	4 GB	Pass=59; Weak=11; Failed=44
$(85,154)[A005408]$	64	4 GB	Pass=54; Weak=14; Failed=46
$(89,210)[A005408]$	64	4 GB	Pass=62; Weak=5; Failed=47
$(30,150)[A005408]$	64	4 GB	Pass=78; Weak=6; Failed=30
$(30,105)[A000045]$	64	4 GB	Pass=82; Weak=2; Failed=30
$(30,150)[A000045]$	64	4 GB	Pass=80; Weak=5; Failed=29

Table 3. NIST test results of the candidate t-NUCAs

t-NUCA	NIST	t-NUCA	NIST	t-NUCA	NIST
(15,180)[A005408]	56/188	(15,210)[A005408]	56/188	(45,166)[A005408]	105/188
(85,154)[A005408]	48/188	(89,170)[A005408]	25/188	(89,210)[A005408]	99/188
(89,240)[A005408]	22/188	(105,154)[A005408]	26/188	(105,166)[A005408]	27/188
(105,180)[A005408]	26/188	(150,210)[A005408]	29/188	(166,75)[A005408]	102/188
(45,15)[A001651]	13/188	(45,75)[A001651]	30/188	(45,154)[A001651]	97/188
(45,166)[A001651]	114/188	(45,170)[A001651]	33/188	(75,89)[A001651]	103/188
(75,154)[A001651]	114/188	(75,166)[A001651]	97/188	(75,170)[A001651]	29/188
(75,180)[A001651]	93/188	(75,210)[A001651]	106/188	(89,45)[A001651]	93/188
(89,75)[A001651]	104/188	(89,154)[A001651]	97/188	(89,166)[A001651]	95/188
(89,180)[A001651]	95/188	(89,210)[A001651]	109/188	(154,45)[A001651]	77/188
(154,166)[A001651]	97/188	(30,150)[A001651]	121/188	(30,45)[A000045]	133/188

Table 4. Some good t-NUCAs based pseudo-random number generators

t-NUCA	NIST	Dieharder
(30,150)[A005408]	135/188	Pass=78; Weak=6; Failed=30
(30,150)[A001651]	121/188	Pass=82; Weak=2; Failed=30
(30,105)[A005408]	103/188	Pass=83; Weak=1; Failed=30
(30,105)[A000045]	137/188	Pass=82; Weak=2; Failed=30
(30,150)[A000045]	147/188	Pass=80; Weak=5; Failed=29
(30,45)[A000045]	133/188	Pass=81; Weak=4; Failed=29
(150,30)[A005408]	159/188	Pass=80; Weak=4; Failed=30
(150,30)[A042968]	164/188	Pass=84; Weak=1; Failed=29
(150,30)[A047201]	160/188	Pass=80; Weak=5; Failed=29
(150,30)[A047253]	157/188	Pass=78; Weak=5; Failed=31

In Table 4, we can find that the t-NUCA $(150, 30)[A042968]$ shows better performance than the other PRNGs. So, this t-NUCA can be a good choice to develop a t-NUCA based PRNG. To verify our claim, we draw space-time diagram of this t-NUCA (see Fig. 7). This figure clearly shows a lack of repeated patterns and seems random, which supports the potential of this window-based system with the CA serving as a good PRNG. In the next section we discuss on development of a output function (temper function) that can enhance the randomness of the proposed CA-based PRNGs.

4   Improving the Proposed PRNG Using Tempering

In the previous section, we discuss about the testing results of the proposed PRNGs. However, there exists no such generator that pass all the tests. In this section, we add an additional output function to enhance the randomness of the proposed t-NUCAs.

4.1   Tempering Function

Here also, we consider 64 bit number as output and for every time, the generator generates next state of the PRNG. But without producing the output directly, like it has been done in Sect. 3, each of the 64 bit numbers is tempered using a tempering function. The state of the generator is updated using its own update rule only after the output has been produced. Here, we use the following tempering function taken from [8]:

$$w = 20 \qquad w = 40 \qquad\qquad w = 100$$

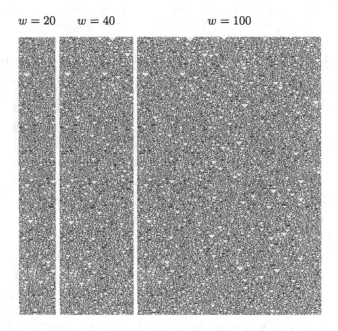

Fig. 7. Space-time diagram of some sample windows for PRNG using a t-NUCA $(150, 30)[A042968]$. Here, the CA size is 500

$$y = x \oplus (x >> u)$$
$$y = y \oplus ((y << s)\ AND\ b)$$
$$y = y \oplus ((y << t)\ AND\ c)$$
$$z = y \oplus (y >> l)$$

Here b and c are appropriate bit-masks of size w, z is the return vector, and u, s, t, and l are integers known as tempering parameters. Shifts, ANDs, ORs, and XORs are the only operations available here. We can assign these variables different initial values and test their efficacy.

Example 6. Let us consider the following values for tempering the output: $u = 3$, $s = 5$, $l = 3$, and $t = 5$. We assign $0XB79B6F16ADD53$ and $0 \times 2EE1E5C1DE43BEF8$ to the bit-marks b and c, respectively. The proposed generator's randomness is improved by using the aforementioned tempering technique. The tempering function's constants have been chosen experimentally for this research, although changing the parameter values can yield better results. The improved report of the NIST and Dieharder tests of some sample t-NUCAs following the use of the tempering function is shown in Table 5.

Table 5. t-NUCAs with tempering function

t-NUCAs	t-NUCA + tempering				
	File size	NIST		Dieharder	
		Passed	Failed	Passed	Weak	Failed
(150,30)[A005408]	1.2 GB	183	5	95	9	10
(150,30)[A042968]	1.2 GB	188	0	96	6	12
(150,30)[A047201]	1.2 GB	188	0	97	5	12
(150,30)[A047253]	1.2 GB	188	0	99	5	10

4.2 Comparison with Other Generators

Using the test suites from NIST and Dieharder, we evaluate the proposed t-NUCA-based generators. For comparison, we consider $MT19937$ (largest period Mersenne Twister) [8], $SFMT19937$ (SIMD-oriented Fast Mersenne Twister) [13], and $TinyMT$ (Tiny Mersenne Twister) [14]. The outcomes are shown in Table 6 and Table 7.

Table 6. Comparison based on Dieharder test suite

File Size	t-NUCA + tempering				Comparisons with other generators					
	(150,30)[A047253]		(150,30)[A047201]		MT		SFMT		TinyMT	
	Weak	Failed	Weak	Failed	Weak	Failed	Weak	Failed	Weak	Failed
12 GB	4	8	4	10	2	0	2	0	3	0
1.2 GB	5	10	5	12	5	3	7	5	5	2

The comparative observations based on Dieharder and NIST are shown in Table 6 and Table 7 respectively. While the proposed PRNGs, that is, $(150, 30)[A042968]$, $(150, 30)[A047253]$, and $(150, 30)[A047201]$ passed all NIST tests (Table 7), they fail a few Dieharder tests. Table 6 shows that their performance is competitive with other existing PRNGs. Hence, t-NUCA is an ideal choice for developing a pseudo-random number generator.

Table 7. Comparison based on NIST test suite

t-NUCA + tempering						Comparisons with other generators			
(150, 30)[A042968]		(150, 30)[A047201]		(150, 30)[A047253]		MT19937-64		SFMT19937-64	
Passed	Failed	Passed	Failed	Passed	Failed	Passed	Failed	Passed	Failed
188	0	188	0	188	0	188	0	188	0

5   Discussion

In this work, we have shown the usability of temporally non-uniform cellular automata (t-NUCAs) in developing a pseudo-random number generator (PRNG). We have addressed the required characteristics that can be inherited in the proposed model to develop a good PRNG. First, we have identified some candidate t-NUCAs that can be useful as pseudo-random number generator. Then we performed empirical tests on the candidate t-NUCAs to identify some good candidates for developing PRNG. Finally we have analyzed the performance of the designed PRNGs over empirical test-beds. However, the list of potential t-NUCAs of developing PRNG is extensive. For further improvement of the work, we need to identify the theory to obtain large cycle of t-NUCA and characterize the t-NUCAs more extensively.

References

1. Bhattacharjee, K., Das, S.: A search for good pseudo-random number generators: survey and empirical studies. Comput. Sci. Rev. **45**, 100471 (2022). https://doi.org/10.1016/j.cosrev.2022.100471
2. Das, S., Roy Chowdhury, D.: CAR30: a new scalable stream cipher with rule 30. Cryptogr. Commun. **5**(2), 137–162 (2013)
3. Kannan, R., Lenstra, A.K., Lovász, L.: Polynomial factorization and nonrandomness of bits of algebraic and some transcendental numbers. In: Proceedings of the Sixteenth Annual ACM Symposium on Theory of Computing, pp. 191–200 (1984)
4. Kowalska, K.A., Fogliano, D., Coello, J.G.: On the revision of nist 800–22 test suites. Cryptology ePrint Archive (2022)
5. L'Ecuyer, P.: Random numbers for simulation. Commun. ACM **33**, 85–97 (1990). https://doi.org/10.1145/84537.84555
6. Marsaglia, G.: Diehard: a battery of tests of randomness. www.stat.fsu.edu/geo (1996)
7. Martinez, G.J.: A note on elementary cellular automata classification. J. Cell. Automata **8**, 233–259 (2013)
8. Matsumoto, M., Nishimura, T.: Mersenne twister: a 623-dimensionally equidistributed uniform pseudo-random number generator. ACM Trans. Model. Comput. Simul. **8**(1), 3–30 (1998)
9. Pal Chaudhuri, P., Roy Chowdhury, D., Nandi, S., Chattopadhyay, S.: Additive cellular automata - theory and applications, vol. 1. IEEE Computer Society Press, USA, ISBN 0-8186-7717-1 (1997)
10. Paul, S., Das, S.: Temporally non-uniform cellular automata (t-NUCA): reversibility and cyclic behavior. Theoretical Computer Science (Submitted) (2023)

11. Paul, S., Roy, S., Das, S.: Pattern classification with temporally stochastic cellular automata. In: Manzoni, L., Mariot, L., Roy Chowdhury, D. (eds.) Cellular Automata and Discrete Complex Systems, AUTOMATA 2023, LNCS, vol. 14152, pp. 137–152. Springer, Cham (2023). https://doi.org/10.1007/978-3-031-42250-8_10

12. Rukhin, A., Soto, J., Nechvatal, J., Smid, M., Barker, E.: A statistical test suite for random and pseudorandom number generators for cryptographic applications. Technical Rep, DTIC Document (2001)

13. Saito, M., Matsumoto, M.: SIMD-oriented fast mersenne twister: a 128-bit pseudorandom number generator. In: Keller, A., Heinrich, S., Niederreiter, H. (eds.) Monte Carlo and Quasi-Monte Carlo Methods 2006, pp. 607–622. Springer, Berlin (2008). https://doi.org/10.1007/978-3-540-74496-2_36

14. Saito, M., Matsumoto, M.: A high quality pseudo random number generator with small internal state. IPSJ SIG Notes 3, 1–6 (2011)

15. Saraniti, M., Goodnick, S.M.: Hybrid fullband cellular automaton/monte carlo approach for fast simulation of charge transport in semiconductors. IEEE Trans. Electron Devices 47(10), 1909–1916 (2000)

16. Sloane, N.J.A.: The on-line encyclopedia of integer sequences® (oeis®). www.oeis.org/ (1964)

A Quantum Walk-Based Scheme for Distributed Searching on Arbitrary Graphs

Mathieu Roget[1,2(✉)] [iD] and Giuseppe Di Molfetta[1,2] [iD]

[1] Laboratoire d'Informatique et Systèmes, Marseille, France
giuseppe.dimolfetta@lis-lab.fr
[2] Université Aix-Merseille, Marseille, France
mathieu.roget@lis-lab.fr

Abstract. A discrete time quantum walk is known to be the single-particle sector of a quantum cellular automaton. Searching in this mathematical framework has interested the community since a long time. However, most results consider spatial search on regular graphs. This work introduces a new quantum walk-based searching scheme, designed to search nodes or edges on arbitrary graphs. As byproduct, such new model allows to generalise quantum cellular automata, usually defined on regular grids, to quantum anonymous networks, allowing a new physics-like mathematical environment for distributed quantum computing.

Keywords: Quantum Distributed Algorithm · Quantum Walk · Quantum Cellular Automata · Quantum Anonymous Network · Searching Algorithm

1 Introduction

Quantum Walks (QW), from a mathematical point of view, coincide with the single-particle sector of quantum cellular automata (QCA), namely a spatially distributed network of local quantum gates. Usually defined on a d-dimensional grid of cell, they are known to be capable of universal computation [1]. Both quantum walks and quantum cellular automata, with their beginnings in digital simulations of fundamental physics [4,6], come into their own in algorithmic search and optimization applications [13]. Searching using QW has been extensively studied in the past decades, with a wide range of applications, including optimisation [14] and machine learning [10]. On the other hand, search algorithms as a field independent of quantum walks have recently been used, as subroutines, to solve distributed computational tasks [7–9]. However, in all these examples, some global information is supposed to be known by each node of the network, such as the size, and usually the network is not anonymous, namely every node has a unique label. Note that, the absence of anonymity and the quantum properties of the algorithm successfully solves the incalculability of certain

M. Dalui et al. (Eds.): ASCAT 2024, CCIS 2021, pp. 72–83, 2024.
https://doi.org/10.1007/978-3-031-56943-2_6

problems such as the leader election problem. However, encoding global information within a quantum state is generally problematic. In order to address such issue, here we first introduce a new QW-based scheme on arbitrary graphs for rephrasing search algorithms. Then we move to the multi-states generalisation, which successfully implement a QCA-based distributed anonymous protocol for searching problems, avoiding any use of global information.

Contribution. Section 2 of this work introduces a Quantum Walk model well suited to search indifferently a node or an edge in arbitrary graphs. In this model the walker's amplitudes are defined onto the graph's edges and ensures 2-dimensional coin everywhere. We detail how this Quantum Walk can be used to search a node or an edge and we show examples of this Quantum Walk on several graphs. In this first part, the Quantum Walk is introduced formally as a discrete dynamical system. In Sect. 3 we move to the multi-particle sector, allowing a many quantum states-dynamics over the network, based on the previous model, and leading to a distributed searching protocol. For the coin and the scattering, local quantum circuits and distributed algorithms are provided. The implementation proposed conserves the graph locality and does not require a node or an edge to have global information like the graph size. The nodes (and edges) do not have unique label, and no leader is needed.

2 Mathematical Model of Quantum Walk on Graphs

This section introduces the quantum walk searching algorithm on undirected connected graphs. First we introduce the quantum walk model on graph and give an example. Then we explain how such a quantum walk can be used to search a marked edge or a marked node. Finally, we show two examples of searching : in the first one we look for an edge and in the second one we look for a node. Finally we derive the asymptotic complexities.

2.1 Quantum Walk Model on Graphs

We consider an undirected connected graph $G = (V, E)$, where V is the set of vertices and E the set of edges. We define the walker's position on the graph's edges and a coin register of dimension two (either $+$ or $-$). The full state of the walker at step t is noted:

$$|\Phi_t\rangle = \sum_{(u,v) \in E} \psi_{u,v}^+ |(u,v)\rangle|+\rangle + \psi_{u,v}^- |(u,v)\rangle|-\rangle.$$

The graph is undirected so we indifferently use $|(u,v)\rangle$ and $|(v,u)\rangle$ to name the same state. Similarly for the complex amplitudes $\psi_{u,v}^+(t) = \psi_{u,v}^-(t)$. We also introduce a polarity for every edge of G. For each of them, we have a polarity function σ, such that : $\forall(u,v) \in E$, $\sigma(u,v) \in \{+,-\}$ and $\sigma(u,v) \neq \sigma(v,u)$. Fig. 1 illustrates how amplitudes and polarity are placed with edge (u,v) of polarity $\sigma(u,v) = 1$.

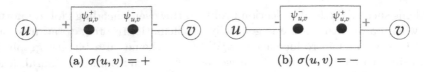

Fig. 1. The edge (u, v) and how are placed the amplitudes for the two possible polarities.

The polarity is a necessary and arbitrary choice made at the algorithm's initialization *independently for every edge*. We discuss why it is necessary and a way to make that choice at the end of this section. The full unitary evolution of the walk reads : $|\Phi_{t+1}\rangle = S \times (I \otimes C) \times |\Phi_t\rangle$, where C is the local coin operation acting on the coin register :

$$\forall (u, v) \in E, \ |(u, v)\rangle|\pm\rangle \xrightarrow{\text{coin}} (I \otimes C) \times |(u, v)\rangle|\pm\rangle = |(u, v)\rangle(C|\pm\rangle).$$

and S is the scattering which moves the complex amplitudes $\alpha_{u,v}^{\pm}$ according to the coin state :

$$\forall u \in V, \ \left(\psi_{u,v}^{\sigma(u,v)}\right)_{v \in V} \xrightarrow{\text{scattering}} D_{\deg(u)} \times \left(\psi_{u,v}^{\sigma(u,v)}\right)_{v \in V},$$

where $D_n = \left(\frac{2}{n}\right)_{i,j} - I_n$.

As an example of the above dynamics, one can consider the path of size 3 with the nodes $\{u, v, w\}$, with polarity $\sigma(u, v) = \sigma(v, w) = +$. Figure 2 shows the unitary evolution of the walker from step t to step $t + 1$, when the coin coincide the first Pauli matrix X.

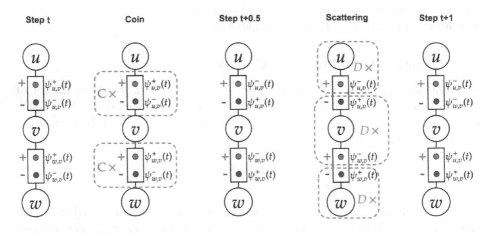

Fig. 2. Example of a walk on a path of size 3.

Polarity : Why and How. There are two reasons we need polarity. First, if the coin operator acts differently on $|+\rangle$ and $|-\rangle$, then different polarities lead to different dynamics. Polarities, in fact, determines how each edge's state scatter within the network, analogously to other spatial searching algorithms on regular lattice [2,3,11,12]. Moreover, polarity is used to divide the edges in two. So each node can only access one amplitude of each edge (depending on the polarity of the edges). This makes things much simpler for a distributed implementation, since we do not have to consider the case of two simultaneous operations on the same amplitude during scattering. One way to initialise the polarity on the graph is to create a coloring of G at the initial time step. Afterward, edges have their $+$ pole in the direction of the node with the higher color. This design is especially convenient for bipartite graphs where every node sees only $+$ polarities or only $-$ polarities.

2.2 Searching

The above dynamical system formally describes a quantum walk on the graph's edges. If the quantum state is measured, we obtain one of the edges of the graph. This scheme is thus well suited to search edges. Thus we may now introduce an oracle (i.e. a black box able to recognize/mark the solution to the searching problem) marking the desired edge $(a, b) \in E$. Analogously with the standard spatial search, the oracle is defined as follows :

$$
\mathcal{O}_f = \left(\sum_{f((u,v))=1} |(u,v)\rangle\langle(u,v)| \right) \otimes R + \left(\sum_{f((u,v))=0} |(u,v)\rangle\langle(u,v)| \right) \otimes I_2,
$$

where f is the classical oracle equals to 1 if and only if the edge is marked, and 0 otherwise. The operator R is a special coin operator which is applied only to the marked edge. Without lack of generality, in the following we set $C = X$ and $R = -X$. The algorithm proposed here is the following:

Algorithm 1. Search a marked edge

Require: $G = (V, E)$ undirected, connected graph
Require: f the classical oracle
Require: $T \in \mathbb{N}$ the hitting time
1: **function** SEARCH(G, f, T)
2:    $|\Phi\rangle \leftarrow \dfrac{1}{\sqrt{2|E|}} \sum_{(u,v)\in E} |(u,v)\rangle|+\rangle + |(u,v)\rangle|-\rangle$ ▷ Diagonal initial state.
3:    **for** $0 \leq i < T$ **do**
4:        $|\Phi\rangle \leftarrow S \times (I \otimes C) \times \mathcal{O}_f \times |\Phi\rangle$ ▷ One step of the walk
5:    **end for**
6:    $(u, v, \pm) \leftarrow$ MEASURE($|\Phi\rangle$)   ▷ Measure the quantum state.
7:    **return** (u, v)
8: **end function**

The initial state is initialized to be diagonal on the basis states and the additional oracle operation is added to the former QW-dynamic. There are two parameters that characterize the above algorithm : the probability of success P of returning the marked edge and the hitting time T. The number of oracle calls of this algorithm is $O(T)$. In practice, we want to choose T such that P is maximal. The previous algorithm can easily be made into a Las Vegas algorithm whose answer is always correct but the running time random. Algorithm 2 shows such transformation. Interestingly, the expected number of times Algorithm 2 calls Algorithm 1 is $O(1/P)$ where P is the probability of success of Algorithm 1. The expected complexity of Algorithm 2 is $O(T/P)$. The complexity of our algorithm is limited by the optimal complexity of $O(\sqrt{K})$ for searching problem where K is the total number of elements. This is the complexity of the Grover algorithm which has been shown to be optimal [16]. However Grover's algorithm assume full connectivity between elements which is not our case. The quantum walk presented here is on the edge of G, which means that the optimal complexity is $O\left(\sqrt{|E|}\right)$.

Algorithm 2. Search a marked edge with guaranteed success

Require: $G = (V, E)$ undirected, connected graph
Require: f, the classical oracle
Require: $T \in \mathbb{N}$ the hitting time
1: **function** GUARANTEEDSEARCH(G, f, T)
2:     $(u, v) \leftarrow (\texttt{nil}, \texttt{nil})$ ▷ Initial value
3:     **while** $f((u, v)) \neq 1$ **do**   ▷ Until we find the marked edge ...
4:         $(u, v) \leftarrow$ SEARCH(G, f, T)   ▷ ... search again.
5:     **end while**
6:     **return** (u, v)
7: **end function**

2.3   Example: Searching an Edge in the Star Graph

In this example we consider the star graph with M edges and $M + 1$ nodes; a graph with $M + 1$ nodes where every node (other than node u_0) is connected to node u_0. Our searching algorithm performs well on this graph, as shown by Theorem 1. In this section, we show the proof of Theorem 1 which is mainly spectral analysis on a simplified dynamic.

Theorem 1. *Let G be the star graph with M edges. Algorithm 1 has an optimal hitting time $T = O\left(\sqrt{M}\right)$ and a probability of success $O(1)$ for G. Algorithm 2 has expected complexity $O\left(\sqrt{M}\right)$.*

Proof. We consider the star graph $G = (V, E)$ of size $M + 1$ such that $V = \{u_0, \dots, u_M\}$ and $E = \{(u_0, u_i) \mid 1 \leq i \leq M\}$. We assume without a loss of generality that the marked edge is (u_0, u_1) and that the polarity is $\forall i > 1$, $\sigma(u_0, u_i) = +$. At any time step t, the state of the walk reads $|\Phi_t\rangle = \sum_{i=1}^{M} \psi_{u_0,u_i}^+(t)|(u_0,u_i)\rangle|+\rangle + \psi_{u_0,u_i}^-(t)|(u_0,u_i)\rangle|-\rangle$. We first show that $\forall t \in \mathbb{N}$, $\forall i > 1$, $\psi_{u_0,u_i}^+(t) = \alpha^+$ and $\psi_{u_0,u_i}^-(t) = \alpha^-$. This greatly simplifies the way we describe the walk dynamic. Afterward we shall provide simple spectral analysis to extract the optimal hitting time T and probability of success P. Next, we prove the property (Q_t) : $\quad \forall i > 1$, $\psi_{u_0,u_i}^+(t) = \alpha_t^+$ and $\psi_{u_0,u_i}^-(t) = \alpha_t^-$ for all $t \in \mathbb{N}$.

The initial state $|\Psi_0\rangle$ is diagonal on the basis states :

$$|\Phi\rangle \leftarrow \frac{1}{\sqrt{2|E|}} \sum_{(u,v)\in E} |(u,v)\rangle|+\rangle + |(u,v)\rangle|-\rangle.$$

All $\left(\psi_{u_0,u_i}^\pm(0)\right)_{i\geq 1}$ are equals so the property Q_0 is satisfied with $\alpha_t^+ = \alpha_t^- = \frac{1}{\sqrt{2M}}$. Now, we assume that Q_t is true and use the walk dynamic to show that Q_{t+1} is also true. The state $|\Psi_{t+1}\rangle$ is described in Table 1 and show that Q_{t+1} is true.

Table 1. Detailed dynamic of the quantum walk for star graphs.

	Step t	After oracle	After coin	After Scattering
α^+	α_t^+	α_t^+	α_t^-	$\alpha_{t+1}^+ = \dfrac{(M-2)\alpha_t^- - 2\psi_{u_0,u_1}^+}{M}(t)$
α^-	α_t^-	α_t^+	α_t^+	$\alpha_{t+1}^- = \alpha_t^+$
ψ_{u_0,u_1}^+	$\psi_{u_0,u_1}^+(t)$	$-\psi_{u_0,u_1}^-(t)$	$-\psi_{u_0,u_1}^-(t)$	$\psi_{u_0,u_1}^+(t+1) = \dfrac{(2M-2)\alpha_t^- + (M-2)\psi_{u_0,u_1}^+}{M}(t)$
ψ_{u_0,u_1}^-	$\psi_{u_0,u_1}^-(t)$	$-\psi_{u_0,u_1}^+(t)$	$-\psi_{u_0,u_1}^+(t)$	$\psi_{u_0,u_1}^-(t+1) = -\psi_{u_0,u_1}^-(t)$

Using the recurrence in Table 1, we can put the dynamic of the walk into a matrix form.

$$\psi_{u_0,u_1}^-(t) = \frac{(-1)}{\sqrt{2M}} \qquad \text{and} \qquad X_{t+1} = AX_t,$$

where

$$X_t = \begin{pmatrix} \alpha_t^+ \\ \alpha_t^- \\ \psi_{u_0,u_1}^+(t) \end{pmatrix} \qquad \text{and} \qquad A = \begin{pmatrix} 0 & \frac{M-2}{M} & \frac{-2}{M} \\ 1 & 0 & 0 \\ 0 & 2\frac{M-1}{M} & \frac{M-2}{M} \end{pmatrix}.$$

After diagonalizing A, the eigenvalues are $\{-1, e^{i\lambda}, e^{-i\lambda}\}$, where $e^{i\lambda} = \frac{M-1+i\sqrt{2M-1}}{M}$. This leads to $\psi_{u_0,u_1}^+(t) \sim \frac{i}{2}\left(e^{-i\lambda t} - e^{i\lambda t}\right) \sim \sin(\lambda t)$, which allows us to deduce the probability p_t of hitting the marked edge $p_t \sim \sin^2(\lambda t)$, and then the optimal hitting time by solving $p_T = 1$:

$$T = \frac{\pi}{2}\frac{1}{\lambda} \sim \frac{\pi}{2}\sqrt{\frac{M}{2}} \sim \frac{\pi}{2\sqrt{2}}\sqrt{M} = O(\sqrt{M}).$$

Finally we have a probability of success $P \sim 1$ for the optimal hitting time $T \sim \frac{\pi}{2\sqrt{2}}\sqrt{M} = O(\sqrt{M})$.

2.4   Searching Nodes

The searching quantum walk presented in the previous section can search one marked edge in a graph. In order to search a node instead, we need to transform the graph we walk on. We call this transformation STARIFY.

Definition 1. STARIFY
Let us consider an undirected graph $G = (V, E)$. The transformation STARIFY *on G returns a graph $\tilde{G}$ for which*

- *every node and edge in G is in $\tilde{G}$ (we call them real nodes and real edges).*
- *for every node $u \in V$, we add a node $\tilde{u}$ (we call these new nodes virtual nodes).*
- *for every node $u \in V$, we add the edge $(u, \tilde{u})$ (we call these new edges virtual edges).*

We call the resulting graph $\tilde{G}$ the starified graph of G.

Searching Nodes. Starifying a graph G allows us to search a marked node u instead of an edge. We can then use the previous searching walk to search the virtual edge and then deduce without ambiguity the marked node u of the initial graph G. This procedure implies that we have to increase the size of the graph (number of edges and nodes). In particular, increasing the number of edges is significant since we increase the dimension of the walker (this can be significant for sparse graphs) and one must be careful of this when computing the complexity. A reassuring result is that searching a node on the complete graph (the strongest possible connectivity) has optimal complexity. As stated in Theorem 2, the complexity is $O(\sqrt{M})$, which is optimal when searching over the edges. Compared to a classical algorithm in $O(M)$ (a depth first search for instance), we do have a quadratic speedup.

Theorem 2. *When using the quantum algorithm to search one marked node in the starified graph $\tilde{G}$ of the complete graph G of size N, the probability of success is $P \sim 1$ and the hitting time $T \sim \frac{\pi}{4}N$.*

Proof. The proof is very similar to the one of Theorem 1. We show that several edges have the same states to reduce the dynamic to a simple recursive equation. We then solve it numerically and derive asymptotic values for the probability of success. A detailed proof shall be provided in the supplementary materials.

3   Distributed Implementation

In this section we move to the multi-particle states dynamics, allowing a distributed implementation of the above searching protocol. We assume that both nodes and edges have quantum and classical registers. They can transmit classical bit and apply controlled operations over the local neighborhood. The nodes need to know their degree but do not require a unique label. We first explain how the quantum registers are dispatched along the graph, then we introduce a distributed protocol to reproduce the dynamic of the walk. In this section we assume that we want to implement the walk on a graph $G = (V, E)$ with polarity σ.

3.1 Positioning the Qubits

The implementation proposed uses W states [5]; i.e. a state of the form $(|001\rangle + |010\rangle + |100\rangle)/\sqrt{3}$. Basically the superposition of all the unary digits. We use two qubits per edge, one for the $+$ state and one for the $-$ state. For every node $u \in V$, we use $\ln(\deg(u)) + 1$ qubits to apply the diffusion across all the neighborhood.

Edges Register The edge register represents the position of the walker and is the one measured at the end of the algorithm. It consists of two qubits per edge, corresponding respectively to the $+$ and $-$ polarity amplitudes. To represent a unary state in this register, we need to introduce the notation $\delta_k^n = \underbrace{0\ldots0}_{k-1 \text{ times}} 1 \underbrace{0\ldots0}_{n-k \text{ times}}$ for the string of n digit with 0 everywhere except at position k where it has 1. For instance, $\delta_2^4 = 0100$. For the sake of simplifying the notations, let us arbitrarily enumerate the edges of E such that we have $E = \{e_1, \ldots, e_{|E|}\}$. The full state of the walk is a linear combination of all the $\left(|\delta_k^{2|E|}\rangle\right)_{k \in 2|E|}$, where $|\delta_{2k}^{2|E|}\rangle$ corresponds to amplitude $|e_k\rangle|+\rangle$ of the mathematical model, and $|\delta_{2k+1}^{2|E|}\rangle$ corresponds to amplitude $|e_k\rangle|-\rangle$. It is important that the global state of the register remains a superposition of $\left(|\delta_k^{2|E|}\rangle\right)_{k \in 2|E|}$, as this allows us to obtain a valid solution to the search problem during measurement. In fact, no matter which state $\delta_k^{2|E|}$ is measured, all the edges are measuring state 0 except one, which is measuring state 1.

Nodes Register. The node register consists of $\log \deg +1$ qubits per node. This register is made up exclusively of auxiliary qubits used for the scattering operation. During the scattering operation, the amplitudes of the edges around a given node u are moved to u's auxiliary qubits. The u's qubits then holds the local amplitudes in binary format, applies the scattering operation and finally transfers these amplitudes back to the local edges. After every step of the quantum walk, the nodes register is in state $|0\ldots0\rangle$ without any entanglement with the edges register.

3.2 Distributed Protocol

The various quantum registers, their sizes, and how they are positioned on the network has been made clear in the previous section. We know give distributed schemes for every operation of the mathematical model: oracle, coin and scattering.

Oracle. In the mathematical model, the oracle applies R on the marked edge only. A definition on the basis of this operator when e_1 is marked would be

$$\begin{pmatrix} |e_1\rangle|+\rangle \\ |e_1\rangle|-\rangle \end{pmatrix} \xmapsto{oracle} R\begin{pmatrix} |e_1\rangle|+\rangle \\ |e_1\rangle|-\rangle \end{pmatrix}.$$

and

$$|e_k\rangle|s\rangle \xmapsto{oracle} |e_k\rangle|s\rangle$$

for all $2 \leq k \leq |E|$, $\forall s = \pm$.

Translated to the distributed registers it means that

$$\begin{pmatrix} |\delta_{2k}^{2|E|}\rangle \\ |\delta_{2k+1}^{2|E|}\rangle \end{pmatrix} \xmapsto{oracle} R\begin{pmatrix} |\delta_{2k}^{2|E|}\rangle \\ |\delta_{2k+1}^{2|E|}\rangle \end{pmatrix}, \qquad \text{and identity everywhere else.}$$

We recall that in our searching algorithm, $R = -X$. This operation can be achieved by two Z gates and a swap applied only to the marked edge.

Coin. The coin operation is very similar to the oracle. We want to apply $C = X$ to all edge states. In the mathematical model, it can be written as

$$\forall k, \ \begin{pmatrix} |e_k\rangle|+\rangle \\ |e_k\rangle|-\rangle \end{pmatrix} \xmapsto{oracle} C\begin{pmatrix} |e_k\rangle|+\rangle \\ |e_k\rangle|-\rangle \end{pmatrix}.$$

This can be realized by a swap applied on all edges.

Scattering. The scattering operation consists in applying a diffusion operator D locally around the nodes. Since scattering around each node is affecting distinct qubits, we only need to design the circuit for one node u of degree d. We note $a_1, \ldots, a_d$ the local edges connected to u. For this section we call η_k the qubits of edge a_k accessible to u following the polarity. Note that, while the local edges (and their qubits) are indexed, this is completely arbitrary and is not changing the final result of the scattering. This nice property is due to the shift invariance of diffusion operator D. The $\log d$ first qubits of u are storing the amplitudes of the local edges in a binary format, while the last one is marking the states we need to act on. We note $k^{(2)}$ the binary representation of k. We define the operator Tr_k between qubits η_k and u register such that

$$|0\rangle|0^{(2)}\rangle|0\rangle \xmapsto{\mathrm{Tr}_k} |0\rangle|0^{(2)}\rangle|0\rangle \qquad \text{and} \qquad |1\rangle|0^{(2)}\rangle|0\rangle \xmapsto{\mathrm{Tr}_k} |0\rangle|(k-1)^{(2)}\rangle|1\rangle.$$

This operator Tr_k admits a circuit depending of k composed of X, CNOT and multi controlled Tofolli gates. Algorithm 3 shows a distributed scheme for realizing this circuit. It is running on the node and uses two communication methods: `RequestCnot`(edge, target) which applies a Not on target controlled by the qubit in edge accessible according to polarity; and `ApplyMCT`(edge) which applies a Not on the edge's qubit controlled by the full node register. In the worst case, this algorithm require $\log d + 1$ calls to `RequestCnot` and one call to `ApplyMCT`. We define Tr by the successive applications of Tr_k between edge a_k and node u for all k. It is worth to note that, while one has to apply every Tr_k successively, the application order does not matter. On the direct basis, the complete scattering operation we want to achieve can be written as

$$\underbrace{|\delta_k^d\rangle}_{\text{edges}} |0^{(2)}\rangle|0\rangle \xrightarrow{\mathrm{Tr}} |0\rangle|k^{(2)}\rangle|1\rangle \xrightarrow{D} |0\rangle D|k^{(2)}\rangle|1\rangle \xrightarrow{\mathrm{Tr}^{-1}} \underbrace{\left(\sum_{j=1}^d D_{k,j}|\delta_j^d\rangle\right)}_{\text{walker scattered}} \otimes|0^{(2)}\rangle|0\rangle.$$

The last qubit allows us to apply D only when the walker was actually on edge k. Following this global idea, Algorithm 4 provides a distributed implementation for the scattering operator. Every node needs a total of $O(d \log d)$ controlled operations (`RequestCnot` or `ApplyMCT`) with the local edges.

Finally, Fig. 3 shows an example of this distributed design of the path graph of five nodes. The path graph has the particularity of having the same topology as a circuit (a qubit being connected to the preceding and following qubit). Notice that such circuit coincide with a partitioned QCA, each operation is local and traslationally invariant.

Algorithm 3. Distributed scheme for Tr_k on node u of degree d

---

Require: e an edge connected to u
Require: $1 \leq k \leq d$
1: **function** $\mathrm{TR}_k(e, k)$
2:      $r \leftarrow \lceil \log d \rceil$ ▷ The size of u's register is $r + 1$
3:      **for** $0 \leq i < r \mid \left((k-1)^{(2)}\right)_i = 1$ **do**
4:          $\mathrm{REQUESTCNOT}(e, q_i)$ ▷ CNOT on the ith qubit of u
5:      **end for**
6:      $\mathrm{REQUESTCNOT}(e, q_r)$
7:      Apply X gate on all qubits q_i of u such that $\left((k-1)^{(2)}\right)_i = 0$.
8:      $\mathrm{APPLYMCT}(e)$
9:      Apply X gate on all qubits q_i of u such that $\left((k-1)^{(2)}\right)_i = 0$.
10: **end function**

Algorithm 4. Distributed scheme for Tr on node u of degree d

Require: $\mathcal{N}$ the set of all edges connected to u
1: **function** $\mathrm{TR}(u)$
2:      $\{a_1, \ldots, a_d\} = \mathcal{N}$ ▷ Random or arbitrary enumeration of the edges
3:      **for** $1 \leq k \leq d$ **do**
4:          $\mathrm{TR}_k(a_k, k)$
5:      **end for**
6: **end function**

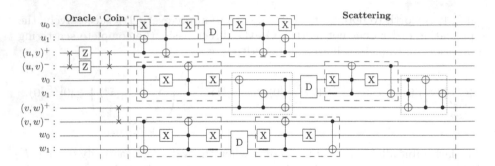

Fig. 3. Circuit of one step of the quantum walk for the path graph $u-v-w$. Dashed lines signal Tr_1 circuit and its inverse while dotted lines Tr_2. The circuit applies successively the oracle on (u,v), the coin, Tr, D, Tr^{-1}.

4   Discussion

The distributed algorithm described in Sect. 3 is based on the quantum walker dynamics introduced earlier for the single particle sector. The most important feature of the above model is that it does not require a leader or unique identifiers. Once the measure is done, all edges measure 0 except one that measure 1, giving us a valid solution; i.e. a solution where every edge agrees on who is marked. The edge who measured 1 can then send the answer to some hypervisor or broadcast it across the network, depending of the context. Notice that this distributed scheme requires that the network be in a W state, which is very common and widely studied in the distributed algorithm community [5,15]. To study the performances of this algorithm, one should use the mathematical model from Sect. 2. In general, it is hard to study analytically this model, except in some rare cases, where a lot of edges have the same state during the whole execution. Indeed, it is mathematically difficult to link the dynamic to the graph structure considering that the latter is on the edges instead of the nodes. In fact, we have to consider a topology based on edges, where two edges are neighbors if they share a node. As such, regular lattices like a grid do not actually have the topology of a grid. However, one could also study this algorithm numerically; preliminary non published work on this aspect have already shown promising results for popular lattices and random graphs. To conclude, it is also important to note that the distributed model recover a QCA-like dynamics on an arbitrary graph. Indeed, in the case of a linear graph, it coincides with the standard partitioned QCA. Overall each operation are node- and edge-independent, so they are translationally invariant. And all operations are local. A byproduct of these properties is a promising distributed architecture which naturally gives a degree of fault tolerance because of local interactions, limiting error propagation.

Aknowledgement. This work is supported by the PEPR EPiQ ANR-22-PETQ-0007, by the ANR JCJC DisQC ANR-22-CE47-0002-01.

References

1. Arrighi, P.: An overview of quantum cellular automata. Nat. Comput. **18**(4), 885–899 (2019)
2. Arrighi, P., Di Molfetta, G., Márquez-Martín, I., Pérez, A.: Dirac equation as a quantum walk over the honeycomb and triangular lattices. Phys. Rev. A **97**(6), 062111 (2018)
3. Bezerra, G., Lugão, P.H., Portugal, R.: Quantum-walk-based search algorithms with multiple marked vertices. Phys. Rev. A **103**(6), 062202 (2021)
4. Bisio, A., D'Ariano, G.M., Perinotti, P.: Quantum walks, deformed relativity and Hopf algebra symmetries. Philos. Trans. R. Soc. A: Math. Phys. Eng. Sci. **374**(2068), 20150232 (2016)
5. Cruz, D., et al.: Efficient quantum algorithms for GHz and w states, and implementation on the IBM quantum computer. Adv. Quantum Technol. **2**(5–6), 1900015 (2019)
6. Di Molfetta, G., Brachet, M., Debbasch, F.: Quantum walks as massless Dirac fermions in curved space-time. Phys. Rev. A **88**(4), 042301 (2013)
7. Gall, F.L., Nishimura, H., Rosmanis, A.: Quantum advantage for the local model in distributed computing. arXiv preprint arXiv:1810.10838 (2018)
8. Izumi, T., Gall, F.L., Magniez, F.: Quantum distributed algorithm for triangle finding in the congest model. arXiv preprint arXiv:1908.11488 (2019)
9. Le Gall, F., Magniez, F.: Sublinear-time quantum computation of the diameter in congest networks. In: Proceedings of the 2018 ACM Symposium on Principles of Distributed Computing, pp. 337–346 (2018)
10. Melnikov, A.A., Fedichkin, L.E., Alodjants, A.: Predicting quantum advantage by quantum walk with convolutional neural networks. New J. Phys. **21**(12), 125002 (2019)
11. Portugal, R.: Quantum Walks and Search Algorithms, vol. 19. Springer, New York (2013). https://doi.org/10.1007/978-1-4614-6336-8
12. Roget, M., Guillet, S., Arrighi, P., Di Molfetta, G.: Grover search as a naturally occurring phenomenon. Phys. Rev. Lett. **124**(18), 180501 (2020)
13. Santha, M.: Quantum walk based search algorithms. In: Agrawal, M., Du, D., Duan, Z., Li, A. (eds.) TAMC 2008. LNCS, vol. 4978, pp. 31–46. Springer, Heidelberg (2008). https://doi.org/10.1007/978-3-540-79228-4_3
14. Slate, N., Matwiejew, E., Marsh, S., Wang, J.B.: Quantum walk-based portfolio optimisation. Quantum **5**, 513 (2021)
15. Tani, S., Kobayashi, H., Matsumoto, K.: Exact quantum algorithms for the leader election problem. ACM Trans. Comput. Theory (TOCT) **4**(1), 1–24 (2012)
16. Zalka, C.: Grover's quantum searching algorithm is optimal. Phys. Rev. A **60**(4), 2746 (1999)

Design of a Cyclone-Proofing Humanity Simulator Using Programmable Cellular Automata

Sudhakar Sahoo[1]([✉]), Swarnava Saha[2], and Abdullah Kazi[3]

[1] Institute of Mathematics and Applications, Bhubaneswar 751029, India
sudhakar.sahoo@gmail.com
[2] Department of Computer Science, The Neotia University, Diamond Harbour Road, Kolkata 743368, West Bengal, India
swarsaha2003@gmail.com
[3] Department of Mechanical Engineering, University of Mumbai, Mahatma Gandhi Road, Mumbai 400032, Maharashtra, India
kazi.abdullah.temea66@gmail.com

Abstract. In this work, the application of Programmable Cellular Automata (PCA) to model and forecast human survival during cyclones is investigated. Using Moore's neighbourhood as the primary framework, the study takes into account important variables including wind speed, population density, time length, and closeness to cyclone pathways. To visualise the effect of different rules on a grid with randomly generated values, a species survival simulator is presented. This study tackles the important question of how humans survive natural disasters and offers insightful information about practical forecasts and mitigation techniques. The study presents a hypothetical use case for modelling cyclone-prone locations, the quantification of parameters, and the CA framework. Visualisations of simulation results are used to showcase the species survival simulator, highlighting its potential to improve resilience and reduce mortality during cyclones. In order to enhance the ecological validity of the simulator and support ecological research and conservation initiatives, future work will integrate real-world cyclone data.

Keywords: Cellular Automata · Survival Prediction · Human Species · Cyclone Effect · Programmable Cellular Automata

1 Introduction

The survival of the human species in the face of natural disasters, particularly cyclones, is a matter of paramount concern. Cyclones, with their destructive forces, have the potential to impact human lives, property, and the environment significantly. Predicting and ensuring the survival of the human population under

This work is carried out as a project in the Summer School on Cellular Automata Technology 2023.

such conditions require innovative approaches that combine computational modeling and an understanding of the complex dynamics of cyclonic events. Different computing paradigm has been used by the researchers. In [16] the integration of advanced machine learning and deep learning models in tropical cyclone prediction has emerged as a vital research frontier. Leveraging meteorological data from MERRA-2 and cyclone data from RSMC - New Delhi, a comprehensive model here employs CNN with optimized hyper parameters using genetic algorithms. This model surpasses traditional classifiers, demonstrating enhanced accuracy, precision, and recall in cyclone categorization. Article [15] contributes insights into Tropical Cyclone Maintenance and Intensification (TCMI) by utilizing the IBTrACS dataset and the MERRA-2 dataset. The study identifies the significance of land-surface processes in accurate TCMI diagnosis, offering a foundation for future analyses. Complementing these efforts, automatic cyclone forecasting through computer vision and satellite image technology is available in [14], achieving accuracy rates of 86% to 95% with a Deep Learning object identification system. In [13], addressing flood submergence modeling challenges, the paper proposes a mathematical model based on bulk method and cellular automata theory, showcasing its utility in designing flood warning systems and emergency plans. Collectively, these studies contribute valuable methodologies and insights, advancing the understanding and prediction of tropical cyclones and related phenomena.

This research focuses on the development of a novel approach to predict and simulate human survival in cyclone-prone areas. Our study leverages the capabilities of Programmable Cellular Automata (PCA) [8–10] as a framework to model and understand the dynamics of human populations during cyclones. While traditional approaches often fall short in accounting for the intricacies of such events, Cellular Automaton (CA) models offer a unique blend of simplicity and the ability to simulate complex behavioral patterns associated with cyclone survival.

A CA model comprises fundamental elements, including cells, states, transition rules, neighborhoods, and a temporal dimension, with transition rules serving as a vital component [1,7,11,12]. These rules capture the mechanisms governing the dynamics of urban systems [2]. In this study, we apply the CA framework to simulate the survival of human populations during cyclones. We draw upon the fundamental principles of cellular automata to model the behavior of individuals in cyclone-prone areas. Our model takes into account various factors, such as population density, Wind speed, and weather conditions, all of which significantly influence human decisions regarding shelter and safety. In a broader context, our work addresses the critical issue of human survival in the face of natural disasters, particularly as climate change continues to impact global weather patterns. With rapid urbanization affecting populations worldwide, our research endeavors to offer insights into effective survival predictions during cyclones.

This paper commences with a general overview of CA and the definition of PCA in Sect. 2, outlining the core principles and their relevance to modeling human survival during cyclones. Section 3 provides a detailed description of our model, Key factors, Different Rules and their Effects, Quantification of Parameters and the Hypothetical use case in this study. Section 4 culminates with simulation results. A research discussion with future work is given in Sect. 5. Section 6 concludes the paper.

2   Cellular Automata

2.1   Definition

In mathematical terms, a Cellular Automaton (CA) can be defined as a quadruple (L, S, N, R), where:

1. L represents the Lattice of cells.
2. S denotes the set of admissible state values.
3. N defines the cell neighborhood in the cell space.
4. R is the transition function, governing state changes.

The framework of a 2D Cellular Automaton serves as a powerful tool for exploring dynamic behaviors in systems where each cells state at position (i, j) is influenced by the transition function R, considering the states of neighboring cells within the defined 2D neighborhood [12].

2.2   Programmable Cellular Automata for Species Survival-Cyclone

1. L (Lattice): L represents the spatial arrangement where the CA simulation takes place. In this context, the lattice corresponds to a 2D grid, and each cell within the grid is considered as a discrete element in the lattice. Each cell, identified by its coordinates (i, j), serves as the fundamental unit within the lattice.
2. S (State Set): S (i, j, t) is the set of all admissible states that each cell can assume during the simulation at time t. In this CA, there are four distinct states:
    - Alive: Representing the state of living individuals.
    - Low Effect: Signifying individuals in a condition of low environmental effect.
    - High Risk: Signifying individuals at a high risk due to surrounding conditions.
    - Dead: Representing the state of deceased individuals.
3. N (Neighborhood): N (i, j, t) defines the spatial relationship among the cells and how they interact with their neighboring cells. In this CA, Moore's Neighborhood is employed, which encompasses eight neighboring cells for each central cell (i, j). These neighbors are:
    - Cells immediately to the north, south, east, and west.
    - Cells diagonally adjacent in the northeast, northwest, southeast, and southwest directions.

4. R (Transition Rules): R defines the rules governing how the state of a cell evolves over time. These rules are applied based on the cell's current state, the states of its neighboring cells, and specific environmental factors. The transition rules are a set of conditions that determine how a cell's state changes in response to the local and global environment. In this CA, transition rules are determined by environmental factors such as wind speed, duration, population density, and proximity to dangerous areas. Based on these factors, the transition rules specify whether a cell remains alive, enters a state of low effect, is at high risk, or becomes deceased in the next time step.

Here we will explain the transition rule what we have used in programming for the purpose of simulation: Let

- windSpeed(i, j), duration(i, j), populationDensity(i, j), and proximity(i, j) be the environmental factors for cell (i, j).

Define the transition rules R based on the quantification of parameters mentioned in Sect. 3.3 as follows:

1. If the following conditions hold for cell (i, j), set the state S(i, j) to alive:

- windSpeed(i, j) ≤ 100 (low wind speed)
- duration(i, j) < 12 h (short duration)
- populationDensity(i, j) ≤ 50 (low population density)
- proximity(i, j) ≥ 10 km (safe proximity)

2. If the following conditions hold for cell (i, j), set the state S(i, j) to low-effect:

- windSpeed(i, j) ≥ 100 (moderate wind speed)
- duration(i, j) ≥ 12 h (medium duration)
- populationDensity(i, j) ≥ 50 (moderate or high population density)
- proximity(i, j) ≤ 5 to 10 km (moderate proximity)

3. If none of the above conditions hold for cell (i, j) and if the following conditions hold, set the state S(i, j) to high-risk:

- windSpeed(i, j) ≥ 100 (moderate wind speed)
- duration(i, j) ≤ 12 h (short duration)
- populationDensity(i, j) ≤ 50 (low population density)
- proximity(i, j) ≤ 10 km (safe proximity)

4. If none of the above conditions hold for cell (i, j), set the state S(i, j) to dead.

These transition rules dictate how the state of each cell in the grid changes based on the environmental conditions specified by the factors. The state transitions are discrete and are based on a set of threshold values for the environmental factors E. So we have defined our 5 elements of CA tuple as (L, S, N, R, E) where E is the environmental factor influencing cells in a 2D grid which we named as PCA. Formally, a transition rule for a PCA for the species survival cyclone problem is defined as follows:

S(i, j, t + 1) = F(S(i, j, t), N(i, j, t), Wind Speed(i, j, t), Duration(i, j, t), Population Density(i, j, t), Proximity(i, j, t))

3   Problem Formulation and Solution Strategies

In this research endeavor, a vital objective is to harness the power of cellular automata, a versatile computational model, to predict and analyze the survival dynamics of human and animal species in the face of calamitous events such as cyclones. Utilizing a 2D grid where cells symbolize individuals with states, the study aims to unravel how cyclones and their aftermath impact the distribution and well-being of these populations. By employing CA rules, specifically Moore's neighborhood [5], which considers the influence of eight adjacent cells, the research facilitates the simulation of species interactions and adaptations under diverse environmental conditions [3,4,6]. This innovative approach promises valuable insights into strategies for enhancing resilience and minimizing casualties during calamities, offering a crucial understanding of ecosystem dynamics in times of crisis. Figure 1 shows a flow chart followed to solve the human species survival prediction problem due to the effect of cyclone using PCA as a model. The research begins by defining the study's parameters and objectives. Random data is then generated to serve as the initial conditions for the simulation. The cellular automata framework is implemented to model system interactions, followed by the simulation of scenarios with various rules. The results are analyzed, leading to the refinement of the model and adjustments

Fig. 1. Process flow diagram followed to solve the human species survival prediction problem due to cyclone on using Programmable Cellular Automata (PCA) as a model.

to parameters. The final simulations are conducted, and the data collected is interpreted and summarized to draw meaningful conclusions from the research findings.

3.1 Key Factors

Factors that directly influence the human species survival are identified and enumerated as follows:

1. Competition and scarcity of resources: When resources are few, people must fight for them, which can worsen general health, raise stress levels, and perhaps lower survival rates.
2. Environmental changes: Survival rates and migration patterns can be directly impacted by environmental changes, including variations in temperature, climatic patterns, and the occurrence of natural catastrophes.
3. Climate and habitat appropriateness: Within an organism's native range, survival rates are correlated with the suitability of its environment. The availability of resources, migratory habits, and overall survival may all be impacted by changes in the climate or habitat deterioration.
4. Disease: Survival rates are greater for those with stronger immune systems or, in the case of human populations, better access to healthcare. Additionally, populations with reduced illness prevalence had higher survival rates.
5. Reproduction tactics: Various species have distinct tactics for reproducing, such as multiplying quickly to produce a large number of offspring or reproducing more slowly to focus more on the survival of each individual. Survival rates can be significantly impacted by these tactics, particularly in surroundings that are changing.
6. Evolution and adaptation: Living things that can adjust to their ever-changing surroundings have a higher chance of surviving. Individuals with beneficial features for various circumstances are also more common in populations with increased genetic variety.
7. Predator-prey dynamics: There is a fine balance between the survival rates of predators and prey. The survival of one species can be impacted by changes in its abundance, creating intricate ecological interactions.

3.2 Exploring Different Rules and Their Effects

Next, we conducted experiments with different sets of rules involving the identified parameters to observe their effects on species survival. Each rule set represented a unique combination of wind speed, population density, time duration, and proximity, and we analyzed how species responded to these scenarios. The rules were implemented using Moore's neighbourhood to simulate the interactions between neighboring cells and their influence on species within the grid. Figures 3 to 6 shows structure of PCA rules used to model the survival conditions of organisms in a 20 × 20 grid based on factors such as wind speed, duration, population density, and proximity to dangerous areas. Cells are represented as alive (white), low effect (yellow), high risk (red), or dead (black), showing how species respond to varying environmental conditions.

3.3   Quantification of Parameters

In this section quantification of parameters has been made for the cyclone data. The range of values for different environmental parameters are set by analysing data available in various sources including scientific papers, government websites, and authoritative databases. For mismatch of information We have considered the average of the values for purpose of quantification and simulation.

1. *Wind Speed:* Measure in km/h or m/s using weather monitoring stations. Establish threshold values based on historical data or regional standards as

- Low Wind Speed: < 100 km/h
- Moderate Wind Speed: 100–150 km/h
- High Wind Speed: > 150 km/h

2. *Duration:* Quantify in hours, based on the duration of cyclone impact. We have set threshold values for significant duration that could impact species survival as follows:

- Short Duration: < 12 h
- Medium Duration: 12–24 h
- Long Duration: > 24 h

3. *Population Density:* Calculate as individuals per square kilometer or square mile in cyclone-prone areas. Identify threshold values for high population density that could affect evacuation and response efforts.

- Low Density: < 50 individuals per square kilometer
- Moderate Density: 50–75 individuals per square kilometer
- High Density: > 75 individuals per square kilometer

4. *Proximity to Cyclone Center:* Measure the distance from the cyclone center in kilometers. Establish threshold values to assess the impact of cyclones at varying distances.

- Close Proximity: < 30 km from the cyclone center
- Moderate Proximity: 30–50 km from the cyclone center
- Far Proximity: > 50 km from the cyclone center

To analyze the impact of cyclone-induced calamities on species survival, focusing on four key parameters: wind speed, population density, time duration, and proximity to the cyclones path. These parameters were selected due to their relevance to real-world ecological systems and their potential influence on species resilience in the face of cyclone disasters and its graph is shown in Fig. 2.

3.4   Hypothetical Use Case

Scenario: Cyclone Survival Simulation for a Coastal Town: Imagine we are conducting a study in a coastal town that is frequently affected by cyclones. Our goal is to understand and predict how these cyclones impact, the human populations survival and well-being using PCA. This CA framework enables us to simulate, in a controlled and computationally efficient manner, how different scenarios, such as wind speed, duration, population density, and proximity, can affect the survival and well-being of the human population during a cyclone.

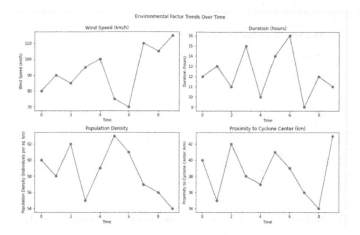

Fig. 2. This graph depicts key environmental factors during a cyclone event. The first subplot shows the changing wind speed (km/h), crucial for assessing cyclone intensity. The second subplot displays cyclone duration (hours), providing insights into its persistence. The third subplot illustrates population density, indicating how it changes over time, affecting human impact assessment. The final subplot shows proximity to the cyclone center (km), influencing cyclone impact intensity. These visualizations help researchers analyze these factors in correlation with a cellular automaton's responses to the cyclone's impact on human survival, facilitating a deeper understanding.

It provides a visual representation of the dynamic changes in the populations status during and after the cyclones impact. By utilizing this PCA model, we can gain valuable insights into strategies for enhancing resilience, minimizing casualties, and improving disaster preparedness in real-world scenarios, which is crucial.

4   Results

The Species Survival Simulator is a web-based simulation designed to model the survival conditions of a species within its habitat. It uses a grid-based representation of the environment and factors in various environmental conditions to determine the health and survival status of the simulated species. With an initial condition of 40% black cells (representing dead cells) and 60% white cells (alive cells), the simulation will maintain the state of the initially "dead" cells, meaning they will not become "alive" again since dead cells cannot be revived. The "alive" cells will potentially change to other states (low-effect, high-risk, or remain alive) based on random factors like wind speed, duration, population density, and proximity to dangerous areas. The exact outcome in terms of the percentage of each state after a certain number of iterations can vary due to the randomness in the simulation. However, it's likely that a portion of the initially "alive" cells will transition to different states, and the final distribution of cell states will depend on the specific factors affecting each cell.

Fig. 3. Initial Configuration of the Species Survival Simulator.

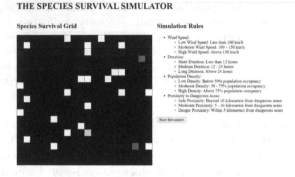

Fig. 4. An intermediate output-1 of the Species Survival Simulator.

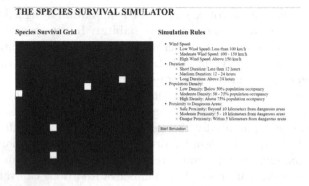

Fig. 5. An intermediate output-2 of the Species Survival Simulator.

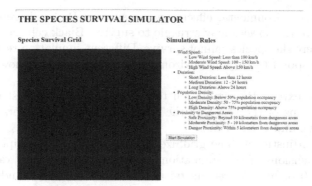

Fig. 6. Output of Species Survival Simulator when all the people are dead as the black cell signifies that and at this point one can stop the PCA based simulation.

4.1 Program Structure

1. HTML Structure: The HTML document defines the structure of the program, including grid visualization and simulation rules. It consists of a grid container and a rules container.
2. CSS Styling: CSS styles are applied to format the grid cells and the rules section. Different cell states are represented by distinct colors, including white (alive), yellow (low effect), red (high risk), and black (dead).
3. JavaScript Logic: JavaScript is used to create and manage the grid, apply simulation rules, and update cell colors based on environmental factors.

4.2 Simulation Process

1. Grid Initialization: The program starts by creating a 20×20 grid representing the habitat of the species. All grid cells are initially set as alive (white).
2. Environmental Factors: The simulation considers four main environmental factors such as wind speed, duration, population density, and proximity to dangerous areas. Thresholds are defined for these factors to determine their impact on the species survival.
3. Applying Simulation Rules: When the Start Simulation button is clicked, the simulation begins. For each cell in the grid, random values are generated for the environmental factors. The average values of these factors are calculated based on neighboring cells. Cell colors are assigned based on the evaluated factors and defined thresholds. The colors can represent low effect, high risk, alive, or dead.

4.3 Simulation Output

Upon starting the simulation, the grid cells colors will change dynamically. White cells represent areas where the species is alive and thriving. Yellow cells indicate areas with low environmental effects on the species. Red cells represent high-risk areas where the species may struggle to survive. Black cells signify extreme conditions where the species cannot survive. Different simulation results showing how species respond to varying environmental conditions are shown in Figs. 3, 4, 5 and 6.

In order to extend the simulation to a larger scale, one must consider the spatial dimensions of the grid and the corresponding adjustments in both the JavaScript logic and CSS styling.

Grid Size Adjustment: The gridSize variable in the JavaScript code determines the dimensions of the simulation grid. By increasing this value, such as setting it to 50 or any desired magnitude, the simulation expands to cover a larger area. This adjustment implies a finer granularity in the spatial representation, offering a more detailed observation of risk factors across the expanded terrain.

const gridSize = 50; /* Adjust to your desired size */

Styling Considerations: Accompanying the grid size adjustment, it is imperative to align the visual representation in the CSS styling. Specifically, the dimensions of each grid cell, governed by the .cell class, should be modified to maintain visual coherence. For example:

.cell

width: 15px; /* Adjust as needed */

height: 15px; /* Adjust as needed */

border: 1px solid black;

In Figs. 7, 8, 9, 10 and 11 we have provided a result for grid size of 100 × 100 with initial cell state to be white in colour with different parameters based on rules resulting in colour shifts. The 5 figures describe the space-time diagrams of species survival simulation in consecutive iterations.

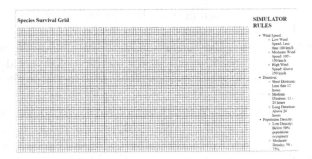

Fig. 7. Initial configuration of the Species Survival Simulator over a grid size of 100 × 100.

In the analysis of cellular automata (CA) rules applied to seemingly identical grids of 100 × 100, the discerning factor lies in the intricate observation of state movements within the grid. These dynamics are crucial for understanding the detailed behavior of the system over time.

Fig. 8. Cyclone effect dynamics of the PCA rule after first iteration.

Fig. 9. Cyclone effect dynamics of the PCA rule after second iteration.

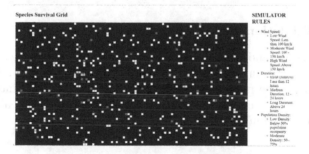

Fig. 10. Cyclone effect dynamics of the PCA rule after third iteration.

Fig. 11. Cyclone effect dynamics of the PCA rule after fourth iteration.

So at the end we want to convey that the theoretical extension of the simulation to a larger scale involves adjusting the grid size, aligning visual representations through CSS styling, and carefully considering the performance implications of the chosen scale. This approach ensures a holistic and meaningful exploration of risk factors across a more extensive spatial context.

5   Discussion and Future Work

This research paper outlines the future work planned to improve the species survival simulator, which employs cellular automata rule with Moore's neighbourhood. Currently, the simulator relies on randomly generated values for cell allocation, but to enhance its ecological validity, real-life data from cyclones is required to be collected and integrated. The aim is to create simulations that accurately reflect the impact of calamities on species survival.

5.1   Data Collection Efforts

In pursuit of authentic data, various sources will be explored, including scientific papers, government websites, and authoritative databases. Relevant variables such as wind direction, wind speed, and human population density from historical cyclones needs to be acquired to represent the conditions experienced during these natural disasters.

5.2   Data Validation and Organization

Rigorous validation and verification processes are required to be undertaken in future to ensure the quality and accuracy of the collected data. The acquired data is required to be meticulously organized to prepare it for integration into the simulator. Proper categorization and formatting are essential to maintain the integrity of the simulation.

5.3   Simulator Enhancement

Once the real-life data is gathered and organized, it will be seamlessly incorporated into the species survival simulator. This integration will elevate the simulator's realism and practical utility in predicting species responses to cyclones.

6   Conclusion

The ongoing efforts to integrate real-life cyclone data into the species survival simulator hold the promise of significantly improving its ecological validity. This enhanced simulator will serve as a valuable tool for ecological research and conservation efforts. The forthcoming research will disseminate the findings, further enriching the scientific community's understanding of species survival dynamics.

This future work seeks to make a meaningful impact on ecological studies and aid in the development of effective strategies to safeguard species in the face of natural disasters and other identified key factors. We will conduct sensitivity analysis by varying the assumptions and rules to see how different factors impact the outcomes. This will help us understand the robustness of our model and the relative importance of different variables. While this theoretical study won't replace real-world data, it can provide a valuable framework for understanding the interplay of different factors and their implications for survival.

Acknowledgement. This work is carried out as a project in the Summer School on Cellular Automata Technology 2023. The authors are grateful to Prof. Sukanta Das and other mentors of the school for their valuable comments and suggestions.

References

1. Adamatzky, A.: Designing Beauty: The Art of Cellular Automata. ECC, vol. 20, pp. 173–179. Springer, Cham (2016). https://doi.org/10.1007/978-3-319-27270-2
2. Almeida, C.M., Gleriani, J.M., Castejon, E.F., Soares, B.S.: Using neural networks and cellular automata for modelling intra-urban land-use dynamics. Int. J. Geogr. Inf. Sci. **22**(9), 943–963 (2008)
3. Balzter, H., Braun, P.W., Kohler, W.: Cellular automata models for vegetation dynamics. Ecol. Model. **107**(2–3), 113–125 (1998)
4. Cluckie, I.D., Rico-Ramirez, M.A., Xuan, Y.: An experiment of rainfall prediction over the Odra catchment by combining weather radar and numerical weather model. In: Proceedings of 7th International conference on Hydroinformatics HIC 2006, Nice, France (2006)
5. Colasanti, R.L., Hunt, R., Watrud, L.: A simple cellular automaton model for high-level vegetation dynamics. Ecol. Model. **203**(3–4), 363–374 (2007)
6. Golding, B.: Nimrod: a system for generating automated very short range forecast. Meteorol. Appl. **5**(01), 1–16 (1998)
7. Ilachinski, A., Cellular Automata : A Discrete Universe. World Scientific, Singapore; River Edge, NJ, xxxii, 808p., (2001)
8. Pal, S., Sahoo, S., Nayak, B.K.: Construction of one-dimensional nonuniform number conserving elementary cellular automata rules. Int. J. Bifurc. Chaos **31**(5), 2150072 (2016)
9. Pal, S., Sahoo, S., Nayak, B.K.: Deterministic computing techniques for perfect density classification. Int. J. Bifurc. Chaos **29**(5), 1950064:1-1950064:11 (2019)
10. Sahoo, S., Pal Choudhury, P., Pal, A., Nayak, B.K.: Solutions on 1-D and 2-D density classification problem using programmable cellular automata. J. Cell. Autom. **9**(1), 59–88 (2014)
11. Wolfram, S.: Cellular automata and complexity. In: Collected Papers, pp. 5–7. Addison-Wesley Publishing Company, Reading, Massachusetts, Menlo Park, CA (1994)
12. Wolfram, S.: A New Kind of Science. Wolfram Media Inc., Champaign (2002)
13. Xi, C.: Cellular automaton model of flood submergence based on bulk method. Environ. Sci. Eng. J. Hydroelect. Eng. **32**, 30–34 (2013)
14. Kavitha, D., Nikhil, T., Praveen, R., Nhj, M.: Detection of tropical cyclone using deep learning network. In: Proceedings of the International Conference on Innovative Computing and Communication (ICICC) (2022)

15. Thomas, A.M., Shepherd, J.M.: A machine-learning based tool for diagnosing inland tropical cyclone maintenance or intensification events. Front. Earth. Sci. Sec. Atmos. Sci. **10**, 818671 (2022)
16. Varalakshmi, P., Vasumathi, N., Venkatesan, R.: Tropical cyclone prediction based on multi-model fusion across Indian coastal region. Progr. Oceanogr. **193**, 1025537 (2021)

Zero Logic Based Stable Three Input QCA XOR Gate

Mrinal Goswami[✉], Avayjeet Paul, and Arpita Nath Boruah

Faculty of Engineering, Assam down town University, Sankar Madhab Path,
Gandhi Nagar, Panikhaiti, Guwahati, Assam, India
{mrinal.g,arpita.b}@adtu.in

Abstract. Quantum-dot Cellular Automata (QCA) is a majority voter based nanotechnology which uses quantum dots as the fundamental building blocks for digital circuits. QCA leverages the principles of quantum mechanics to perform logic operations and store information. QCA has the advantage of low power consumption because it primarily relies on the electrostatic interactions of electrons in quantum dots, reducing energy dissipation compared to traditional transistor technology. Traditionally QCA gates are realized using 3 input majority voter and inverter. However, in this work, a three input cell interaction based (Zero Logic) QCA XOR gate is presented. Physical verification of proposed zero logic structure has done and found stable. The presented zero logic design is further tested designing a full adder. Simulation outcome shows its effectiveness in terms of all the acceptable parameters of QCA simulation. QCADesigner version 2.0.3 is employed to simulate all the proposed designs considering all default parameters in bistable approximation.

Keywords: Quantum dot Cellular Automata · XOR Gate · Zero Logic · QCA Cell Interaction · Cellular Automata

1 Introduction

Quantum-dot cellular automata (QCA) is a promising nanotechnology-based approach for building digital circuits at the molecular or quantum level. It's an alternative to traditional semiconductor-based computing. QCA relies on the behavior of quantum dots, which are nanoscale semiconductor structures that can trap and release electrons, effectively representing binary information as the absence or presence of an electron [11]. A QCA cell is made up of two electrons and four quantum dots. The coulombic interaction between the cells is used to transmit information from one QCA cell to another. The radius of impact, which controls how far each QCA cell will interact with its neighbours, is essential in this situation.

The fundamental gates employed in QCA technology are the three-input majority voter and inverter. Numerous studies have been done to construct QCA circuits using basic primitives (Three Input Majority Voter and Inverter)

M. Dalui et al. (Eds.): ASCAT 2024, CCIS 2021, pp. 99–108, 2024.
https://doi.org/10.1007/978-3-031-56943-2_8

[1,4,8,10,16,17]. However, all of the prior designs used the same traditional gate-level implementation. In [14,15], universal logic gates (CMVMIN) are presented so that majority and minority voters can be produced from the same circuit.

Exclusive OR (XOR) gates have many different uses in digital circuits [9]. XOR gate may be used for modulo-2 operations and also ALUs employ it for binary additions. A crucial use of an XOR circuit is in the search for a certain bit pattern in a very long data sequence because of the property that an XOR circuit emits a zero when both of its inputs match. Additionally, some circuits employ XOR gates as linear feedback shift registers to produce pseudorandom numbers. Few attempts have been made recently to realise XOR gates utilising explicit QCA cell contact [2,3,5,6,13,19]. These type of QCA circuits do not follow any logic. The adder performance can be greatly enhanced by the gate level of the cell interaction QCA-based XOR gate [3].

The above background encourage us to realize a novel XOR gate which can serve all the requirements of QCA design. This research's main contribution can be summed up as follows:

- A zero logic novel XOR gate is proposed and simulated.
- The stability of electrons are investigated and compared with existing designs.
- Higher level logic circuits with superior outcomes in terms of area, cell count, delay and cost.

The structure of this paper is as follows: The background of QCA and a review of earlier XOR work are presented in Sects. 2 and 3 respectively. In Sect. 4, the unique XOR logic gate's design, layout, and analyses (simulation, stability checking, power consumption, reliability, etc.) are introduced. Section 5 contains the conclusion.

2   Principles of QCA

Quantum-dot Cellular Automata (QCA) is a novel and emerging nanotechnology-based computing paradigm that utilizes the principles of quantum mechanics to perform digital logic operations and store information. A QCA cell consist of four quantum dots where maximum two electrons can move from one dot to another as shown in Fig. 1(a). The movement of these electrons can be in any directions however, for stability they actually reside only in two positions as shown in Fig. 1(b). These positions are treated as positive 1 and negative 1 according to their position of stability. As we already mentioned, QCA is a majority voter based nanotechnology. The commonly used majority voter in QCA is 3 input majority voter ($MV(E,F,G) = EF + FG + GE$ where E, F, and G are its three inputs) which is shown in Fig. 1(c). The inversion operation can be performed with the help of inverter as shown in Fig. 1(d). QCA operates in a clocked fashion as shown in Fig. 1(e). A global clock signal synchronizes the switching of electrons within the quantum dots. There are four types of clock zones used in QCA: clock zone 0, clock zone 1, clock zone 2 and clock zone 3.

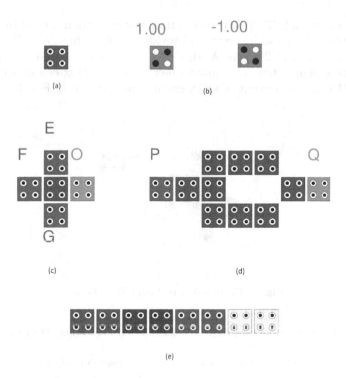

Fig. 1. QCA Basics

3   Related Work

Three input QCA XOR gate design mainly can be distributed into two types: (i) Logic Based (Traditional Approach) and (ii) Zero Logic Based (Cell Interaction based approach). In QCA, logic is developed and realized with the help of three input majority voter and inverter. However, it is found that the structure with more majority gates has a higher chance of getting error [19]. In [3], authors suggest that the performance of higher order circuits are greatly enhanced if it is realized with zero logic based building blocks. However, infancy work has been found in this regard [2,3,5,6,19]. Most of these existing designs are not fabrication friendly as most of these designs uses rotating QCA cell which is difficult to produce [9]. Also, many of these designs needs translation of QCA cells to produce the desired output. Moreover, existing zero logic based designs cover high area, delay and cell count.

4   Proposed Design

Three-input majority gates and inverters are the basic logic devices utilised in QCA. Contrary to usual design methodologies, the proposed gate uses a zero logic design approach (Cell Interaction Based) rather than a combination of 3-input majority gates and inverters to provide the XOR function. The proposed design is shown in Fig. 2 where A, B, C will serve as input and O will serve as output. The design is free from any wire-crossing, rotating cell and translation of any QCA cell. The output of QCA simulation is shown in Fig. 3.

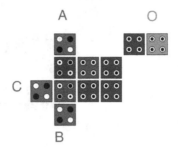

Fig. 2. Proposed Zero Logic XOR Gate

Table 1. Performance Analysis of Proposed Zero Logic XOR Gate

Design	Cell Count	Area (nm^2)	Clock Zone	Cost	Translation Required	Rotation Required
In [2]	14	16284	2	1	Not required	Not required
In [6]	14	11564	2	3	Not required	Not required
In [5]	14	13688	2	3	Required	Not required
In [3]	12	11564	2	1.5	Not required	Not required
Proposed	11	9204	2	1	Not required	Not required

According to [12], total count of QCA gates, delay, and wire crossings are the three key metrics that must be considered in order to calculate the cost of QCA designs. The function to estimate cost of any QCA circuit is proposed in [12] which is as follows: $Cost = (P^M + Q + R^N) \times S^O, 1 \leq M, N, O$ where P represents majority gate count, Q is the inverter count, R is the crossovers count, S is the delay of the circuit, and M, N, O is the exponential weightings for majority gate count, wire crossover count and delay respectively. The values of M, N and O in this work are fixed at 1, 1, and 1, respectively. The performance comparison of the zero logic XOR gate and all other previous gate is shown in Table 1. It is evident that the proposed logic is serving better in all the parameters of QCA simulation. Moreover, with no translation and rotation it is also served as a good candidate for fabrication friendly design.

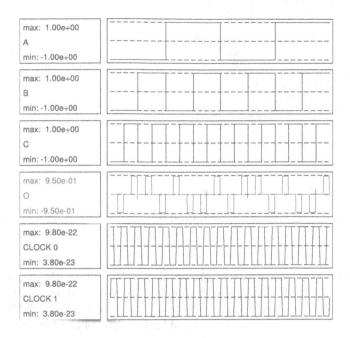

Fig. 3. Output of Zero Logic XOR Gate

4.1 Stability Checking

The physical verification of cell layout is described in this part in order to assess the stability of the proposed XOR architecture. Each QCA cell is 18×18 nm in dimension, and there is a 2 nm gap between adjacent QCA cells. Additionally, it is assumed that the input cell's polarisation is A = B = C = 1 as illustrated in Fig. 4. The calculated kink energy of the output cell is used to examine the stability of the proposed XOR gate. To achieve the stability of any QCA system, the kink energy should be as low as possible. The Eq. 1 is used to compute the kink energy between two electron charges, where U is the kink energy, k is a fixed constant, q_1 and q_2 are electric charges, and r is the distance between the two charges. We enter the values of k and q in Eq. 1 to get Eq. 2.

$$U = \frac{kq_1q_2}{r} \tag{1}$$

$$kq_1q_2 = 9 \times 10^9 \times (1.6)^2 \times 10^{-38} \tag{2}$$

According to Eq. 2, the kink energy between the electrons (1 to 14) with regard to electrons ("X") and ("Y") for Fig. 4(a) is listed in Table 2. The same process is repeated for Fig. 4(b) and found $4.43 \times 10^{-20} J$ and $17.54 \times 10^{-20} J$ for "X" and "Y" electron respectively. According to [7], the structure shown in Fig. 4(a)

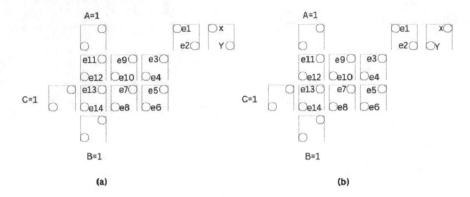

Fig. 4. Position of electrons on the proposed XOR gate(a) XOR with O =-1 (b) XOR with O= 1

is more stable than the structure shown in Fig. 4(b). Recently, some zero logic design of configurable XOR/XNOR function is proposed in [3,19]. However, the physical stability found in [3,19] are $32.41 \times 10^{-20} J$ and $25.92 \times 10^{-20} J$ respectively. On the other hand, the total kink energy of the proposed zero logic structure is estimated as $12.73 \times 10^{-20} J$.

Table 2. Estimation of the kink energy at the proposed XOR Gate's x and y positions(Fig. 4(a))

For X-electron	For Y-electron
$U_1 = \frac{23.04 \times 10^{-29}}{20 \times 10^{-09}} = 1.15 \times 10^{-20} J$	$U_1 = \frac{23.04 \times 10^{-29}}{42.05 \times 10^{-09}} = 0.54 \times 10^{-20} J$
$U_2 = \frac{23.04 \times 10^{-29}}{18.11 \times 10^{-09}} = 1.27 \times 10^{-20} J$	$U_2 = \frac{23.04 \times 10^{-29}}{20 \times 10^{-09}} = 1.15 \times 10^{-20} J$
$U_3 = \frac{23.04 \times 10^{-29}}{29.73 \times 10^{-09}} = 0.77 \times 10^{-20} J$	$U_3 = \frac{23.04 \times 10^{-29}}{40.05 \times 10^{-09}} = 0.57 \times 10^{-20} J$
$U_4 = \frac{23.04 \times 10^{-29}}{55.17 \times 10^{-09}} = 0.41 \times 10^{-20} J$	$U_4 = \frac{23.04 \times 10^{-29}}{64.40 \times 10^{-09}} = 0.35 \times 10^{-20} J$
$U_5 = \frac{23.04 \times 10^{-29}}{45.65 \times 10^{-09}} = 0.50 \times 10^{-20} J$	$U_5 = \frac{23.04 \times 10^{-29}}{45.65 \times 10^{-09}} = 0.50 \times 10^{-20} J$
$U_6 = \frac{23.04 \times 10^{-29}}{70.46 \times 10^{-09}} = 0.32 \times 10^{-20} J$	$U_6 = \frac{23.04 \times 10^{-29}}{70.46 \times 10^{-09}} = 0.32 \times 10^{-20} J$
$U_7 = \frac{23.04 \times 10^{-29}}{58.28 \times 10^{-09}} = 0.39 \times 10^{-20} J$	$U_7 = \frac{23.04 \times 10^{-29}}{63.91 \times 10^{-09}} = 0.36 \times 10^{-20} J$
$U_8 = \frac{23.04 \times 10^{-29}}{83.45 \times 10^{-09}} = 0.27 \times 10^{-20} J$	$U_8 = \frac{23.04 \times 10^{-29}}{87.66 \times 10^{-09}} = 0.26 \times 10^{-20} J$
$U_9 = \frac{23.04 \times 10^{-29}}{46.52 \times 10^{-09}} = 0.49 \times 10^{-20} J$	$U_9 = \frac{23.04 \times 10^{-29}}{60.03 \times 10^{-09}} = 0.38 \times 10^{-20} J$
$U_{10} = \frac{23.04 \times 10^{-29}}{55.17 \times 10^{-09}} = 0.41 \times 10^{-20} J$	$U_{10} = \frac{23.04 \times 10^{-29}}{80.51 \times 10^{-09}} = 0.28 \times 10^{-20} J$
$U_{11} = \frac{23.04 \times 10^{-29}}{71.02 \times 10^{-09}} = 0.32 \times 10^{-20} J$	$U_{11} = \frac{23.04 \times 10^{-29}}{80.02 \times 10^{-09}} = 0.28 \times 10^{-20} J$
$U_{12} = \frac{23.04 \times 10^{-29}}{65.15 \times 10^{-09}} = 0.35 \times 10^{-20} J$	$U_{12} = \frac{23.04 \times 10^{-29}}{100.02 \times 10^{-09}} = 0.23 \times 10^{-20} J$
$U_{13} = \frac{23.04 \times 10^{-29}}{88.57 \times 10^{-09}} = 0.26 \times 10^{-20} J$	$U_{13} = \frac{23.04 \times 10^{-29}}{82.97 \times 10^{-09}} = 0.27 \times 10^{-20} J$
$U_{14} = \frac{23.04 \times 10^{-29}}{98.81 \times 10^{-09}} = 0.23 \times 10^{-20} J$	$U_{14} = \frac{23.04 \times 10^{-29}}{105.85 \times 10^{-09}} = 0.21 \times 10^{-20} J$
$U_{T1} = 7.03 \times 10^{-20} J$	$U_{T2} = 5.70 \times 10^{-20} J$

4.2 Higher Order Designs

The scalability and flexibility of the proposed zero logic XOR gate is further tested by building a magnitude comparator and a full adder circuit, discussed below.

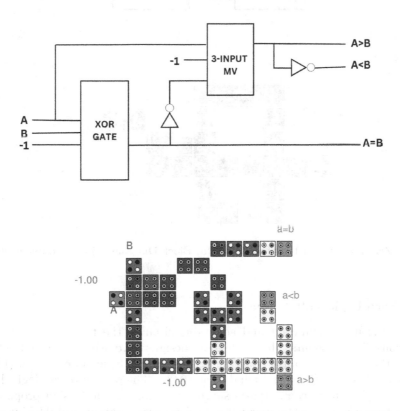

Fig. 5. Zero Logic based Magnitude Comparator (a) Block Diagram (b) QCA Implementation

4.3 Zero Logic Comparator

Using a comparator, two numbers, A and B, are compared in terms of their respective magnitudes. The comparison output is specified by three binary variables. For a one bit comparator, all that is required to compare two binary numbers is their first bit. The proposed one bit comparator can be seen in Fig. 5. The proposed zero logic comparator covered an area of 0.04 μm^2 along with a delay of 1. The same zero logic XOR gate is employed to design the circuit.

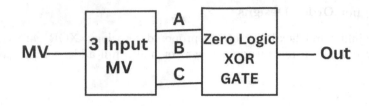

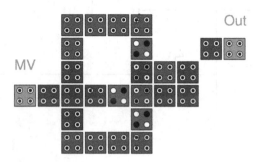

Fig. 6. Zero Logic based 1-bit Full Adder (a) Block Diagram (b) QCA Implementation

4.4 Zero Logic Full Adder

One of the most significant and adaptable circuits, the full adder, is used in numerous digital circuits. Many full adder designs have been researched in QCA due to their immense value. Here, we suggest a zero logic adder design (Fig. 6) with a modest area (0.03 μm^2) and delay (0.05). The proposed zero logic design is compared with its counterparts suggested in [3,19]. The design proposed in [19] utilizes 0.75 delay, 60 QCA normal cells, covering an area of 0.057 μm^2. In [3], the structure utilizes 0.75 delay, 14 QCA normal cells, covering an area of 0.016 μm^2. QCADesigner version 2.0.3 is employed to simulate all the proposed designs discussed in this article considering all default parameters in bistable approximation [18].

5 Conclusion

This article presents a zero logic based three input XOR gate. Unlikely to traditional design approach, the proposed design is based on cell interaction. The XOR gate is compared with all the available standard parameters of QCA design and found most efficient. The proposed design is taking less area compared to all the previous design of zero logic based. The stability of the proposed design is tested and found stable compared to all the previous zero logic based XOR design. The flexibility of the proposed design is covered building higher order circuits such as full adder and comparator. The full adder proposed here is the most compact design considering all the previous adder proposed in literature.

References

1. Ahmad, F., din Bhat, G.M.: Design of novel inverter and buffer in quantum-dot cellular automata (QCA). In: 2015 2nd International Conference on Computing for Sustainable Global Development (INDIACom), pp. 67–72 (2015)
2. Ahmad, F., Bhat, G.M., Khademolhosseini, H., Azimi, S., Angizi, S., Navi, K.: Towards single layer quantum-dot cellular automata adders based on explicit interaction of cells. J. Comput. Sci. **16**, 8–15 (2016)
3. Ahmadpour, S.S., Navimipour, N.J., Mosleh, M., Bahar, A.N., Yalcin, S.: A nanoscale n-bit ripple carry adder using an optimized XOR gate and quantum-dots technology with diminished cells and power dissipation. Nano Commun. Netw. **36**, 100442 (2023)
4. Angizi, S., Alkaldy, E., Bagherzadeh, N., Navi, K.: Novel robust single layer wire crossing approach for exclusive or sum of products logic design with quantum-dot cellular automata. J. Low Power Electron. **10**(2), 259–271 (2014)
5. Balali, M., Rezai, A., Balali, H., Rabiei, F., Emadi, S.: Towards coplanar quantum-dot cellular automata adders based on efficient three-input XOR gate. Results Phys. **7**, 1389–1395 (2017). https://doi.org/10.1016/j.rinp.2017.04.005
6. Chabi, A.M., et al.: Towards ultra-efficient QCA reversible circuits. Microprocess. Microsyst. **49**, 127–138 (2017)
7. Farazkish, R.: A new quantum-dot cellular automata fault-tolerant full-adder. J. Comput. Electron. **14**(2), 506–514 (2015). https://doi.org/10.1007/s10825-015-0668-2
8. Goswami, M., Kumar, B., Tibrewal, H., Mazumdar, S.: Efficient realization of digital logic circuit using QCA multiplexer. In: 2014 2nd International Conference on Business and Information Management (ICBIM), pp. 165–170 (2014)
9. Goswami, M., Mohit, K., Sen, B.: Cost effective realization of XOR logic in QCA. In: 2017 7th International Symposium on Embedded Computing and System Design (ISED). pp. 1–5 (2017). https://doi.org/10.1109/ISED.2017.8303950
10. Hayati, M., Rezaei, A.: Design of novel efficient XOR gates for quantum-dot cellular automata. J. Comput. Theor. Nanosci. **10**(3), 643–647 (2013)
11. Lent, C.S., Tougaw, P.D., Porod, W., Bernstein, G.H.: Quantum cellular automata. Nanotechnology **4**(1), 49 (1993)
12. Liu, W., Lu, L., O'Neill, M., Swartzlander, E.E.: A first step toward cost functions for quantum-dot cellular automata designs. IEEE Trans. Nanotechnol. **13**(3), 476–487 (2014). https://doi.org/10.1109/TNANO.2014.2306754
13. Mahalat, M.H., Goswami, M., Mondal, S., Mondal, A., Sen, B.: Design of fault tolerant nano circuits in QCA using explicit cell interaction. In: 2017 IEEE Calcutta Conference (CALCON), pp. 36–40 (2017). https://doi.org/10.1109/CALCON.2017.8280691
14. Sen, B., Saran, D., Saha, M., Sikdar, B.K.: Synthesis of reversible universal logic around QCA with online testability. In: 2011 International Symposium on Electronic System Design, pp. 236–241 (2011)
15. Sen, B., Sengupta, A., Dalui, M., Sikdar, B.K.: Design of universal logic gate targeting minimum wire-crossings in QCA logic circuit. In: 2010 53rd IEEE International Midwest Symposium on Circuits and Systems, pp. 1181–1184 (2010)

16. Singh, G., Sarin, R.K., Raj, B.: A novel robust exclusive-or function implementation in QCA nanotechnology with energy dissipation analysis. J. Comput. Electron. 15, 455–460 (2016). https://doi.org/10.1007/s10825-016-0804-7
17. Teja, V.C., Polisetti, S., Kasavajjala, S.: QCA based multiplexing of 16 arithmetic 00026; logical subsystems-a paradigm for nano computing. In: 2008 3rd IEEE International Conference on Nano/Micro Engineered and Molecular Systems, pp. 758–763 (2008). https://doi.org/10.1109/NEMS.2008.4484438
18. Walus, K., Dysart, T., Jullien, G.A., Budiman, R.: QCADesigner: a rapid design and simulation tool for quantum-dot cellular automata. Trans. Nanotechnol. 3(1), 26–29 (2004)
19. Wang, L., Xie, G.: A novel XOR/XNOR structure for modular design of QCA circuits. IEEE Trans. Circuits Syst. II Express Briefs 67(12), 3327–3331 (2020). https://doi.org/10.1109/TCSII.2020.2989496

Cellular Automata Based Multiple Stream Parallel Random Number Generator for 64-Bit Computing

Kamalika Bhattacharjee[✉] and Suraj Kumar

Department of Computer Science and Engineering, National Institute of Technology,
Tiruchirappalli 620015, Tamil Nadu, India
kamalika.it@gmail.com

Abstract. Generating random numbers in parallel streams has significant applications in fields like simulation, machine learning, and deep learning. These applications often require the rapid generation of large volumes of random numbers in a way that is reproducible, portable, and efficient. This research specifies an enhanced approach to 64-bit random number generation, aiming to improve upon the standard CPU-based implementations of pseudo-random number generators (PRNGs) such as the Mersenne Twister. The proposed method involves a streamlined version of the algorithm that is designed for parallel execution on multi-core CPUs, with the goal of achieving substantial speed enhancements compared to existing algorithms. To achieve this, this work leverages a 2-state 3-neighborhood maximal length linear cellular automaton as the foundational model for the pseudo-random number generator (PRNG). Notably, the new generator successfully passes almost all benchmark empirical tests for randomness as specified by the NIST and Dieharder test suites and is at par with the Mersenne Twister.

Keywords: Pseudo-Random Number Generators (PRNGs) · Multiple Stream Generator · Maximal Length Cellular Automata · Dieharder · NIST · Multi-core CPUs

1 Introduction

Pseudo-random number generation is a fundamental field in computing focused on creating sequences of seemingly random numbers for applications like Monte Carlo simulations, deep learning, and gaming. These applications demand the rapid and consistent generation of random numbers, a task efficiently handled by a pseudo-random number generator (PRNG). (In this work, by a 'random' number, we mean a 'pseudo-random' number only.) Pseudo-random number generators (PRNGs) operate through an underlying algorithm that begins with a

This work is carried out as a project in the Summer School on Cellular Automata Technology 2023.

random seed and proceeds to generate subsequent numbers in the sequence. In essence, a PRNG is a deterministic computer algorithm that, when given a seed as input, produces a sequence of numbers that mimics randomness and (generally) follows a uniform distribution. However, it is important to note that since the PRNG algorithm is deterministic, it does not produce *truly* random numbers. Instead, the usual goal is to ensure that the PRNG generates numbers that appear random and are uniformly distributed. To assess the quality of a PRNG's output, it must undergo rigorous statistical testing. Test suites like Dieharder, NIST, and TestU01 serve as benchmarks for these tests. When a PRNG performs well in these assessments, it indicates that the generated sequence is statistically random enough for practical applications and possesses a sufficiently large *period*. The period of a PRNG refers to the number of unique outputs it can generate before repeating its initial output, and for practical purposes, it is essential to have a very large period. A d-state PRNG with an internal state of size n is said to have a *maximal* period, if it can generate all numbers except one as part of the same period, that is, its period length is $d^n - 1$.

Over the time, various categories of PRNGs were devised. Examples include Linear Congruential Generators (LCGs), which employ modular arithmetic operations, Linear Feedback Shift Register(LFSR)-based, which use shifts and linear transformations on the previous state vector, and Inversive Congruential Generators (ICGs), which utilize modular multiplicative inverses [1]. Among these, LFSR based generators became most popular as they can directly utilize computer's binary arithmetic for bitwise operations. However, the Mersenne Twister (MT) family of PRNGs, one variant of LFSR-based PRNGs created by Makoto Matsumoto and Takuji Nishimura [2], has stood out as one of the most efficient and widely adopted versions of PRNGs, and the largest version of Mersenne Twister `MT-19937`, have been integrated as the default PRNGs in many popular programming languages like PHP, Ruby, Python, and R. The `MT-19937` boasts a *maximal* period of $2^{19937} - 1$. While the MT family has addressed numerous issues plaguing earlier PRNGs, such as sequential correlation and inadequate period length, there remains room for enhancement. Consequently, a central objective of this paper is to introduce an improved algorithm capable of surpassing the performance of Mersenne Twister algorithms. For this we are going to utilize cellular automata (CAs). This choice is motivated by the historical simplicity and reliability associated with CAs. CAs have long been recognized as a trustworthy method for generating random sequences having their rules well-suited in creating maximal length sequences of random numbers that meet the rigorous criteria of standard randomness tests [1,3].

For applications demanding the simultaneous generation of a large quantity of random numbers, harnessing the formidable computational power of a GPU is a viable strategy. However, this requires the utilization of a lightweight algorithm, rendering existing CPU-based PRNG algorithms ineffective for such a task [4,5]. Moreover, for utilizing the computational power of the multi-core CPUs also, the need is to have a lightweight generator not exceeding 64 bit internal state to exploit current 64-bit computing. In this scenario, the primary challenge lies in developing an algorithm that strikes a balance: it must be robust enough to produce results that successfully pass all benchmark test suites for randomness while also remaining lightweight in implementation, facilitating the creation of multiple parallel streams of random numbers. To tackle this challenge, we are adopting a cellular automaton (CA) based approach, specifically focusing on CA rules 90 and 150, which have traditionally been used to generate 32-bit random numbers [1,6]. Recently, a special CA called $CA(150')$ has been utilized to introduce multiple stream parallel PRNG for 32-bit GPUs [7]. In that work, a 35 length $CA(150')$ is used to generate 32 bit random numbers. It was designed specifically to use the parallelism of 32-bit GPUs, so the CA size is small having the period of the PRNG at maximum $2^{35} - 1$ with 3-bit of wastage per configuration on output generation. However, because of having such a small period, it fails to pass many significant tests in the empirical testbeds. If we focus on the current multi-core 64-bit CPUs, we can increase the internal state to 64 for generating 64-bit numbers that uses the multi-threading of these CPUs. In this paper, our objective is to enhance and adapt this approach to produce 64-bit random numbers effectively. We directly take a 64 length maximal length CA for this purpose such that the PRNG has a maximal period of $2^{64} - 1$ with no wastage of bits in extracting the output numbers.

2   Background

2.1   Cellular Automata

A cellular automaton (CA) is a discrete dynamical system that consists of a regular network of finite state machines, also called *cells*. This work uses a binary nearest neighborhood cellular automaton (also called an elementary CA or ECA) of size n under null boundary condition. Our CA is *non-uniform* – each cell of the CA may use a different rule than the other cells. In the context of an n-cell CA, there are a total of 2^n unique binary configurations that the CA can produce. However, for any $n > 3$ under null boundary condition, at maximum only $2^n - 1$ unique configurations can be generated by a CA starting from an initial configuration. Such a CA is called a *maximal length CA*. Therefore, our goal is to identify a cellular automaton capable of generating this largest possible sequence of numbers in a single cycle, effectively a *maximal length* CA.

For this purpose, we utilize a simple CA which uses elementary CA rule 150 at cell 3rd and 5th while other cells use CA rule 90 as our PRNG:

$$R_{90} = a_{i-1} + a_{i+1} \quad (\text{mod } 2)$$

$$R_{150} = a_{i-1} + a_i + a_{i+1} \quad (\text{mod } 2)$$

Here, a_i, a_{i-1} and a_{i+1} are the present states of cells i, $i-1$ and $i+1$ respectively. Rules 90 and 150 are *linear* rules for CAs that have proven to be successful in producing maximal length CAs [1,8]. It has been demonstrated that, using combination of rule 90 and rule 150, maximal length CA can be constructed for

$n = 64$, that is, a cycle length $2^{64} - 1$ can be generated if we use rule 150 at 3rd and 5th cells while all other cells in the 64-cell CA use CA rule 90 [9]. Such a CA can be characterized by linear algebra by generating its characteristics matrix:

Definition 1. *The characteristic matrix T is a matrix of order $n \times n$ that is defined as follows:*

$$T[i, j] = \begin{cases} 1, & \text{if cell } i \text{ depends on cell } j \\ 0, & \text{otherwise} \end{cases}$$

For example, for a 6-cell CA that uses rule 150 in the fourth and fifth cells and rule 90 in the other four cells, the characteristic matrix would be as follows:

$$T = \begin{bmatrix} 0 & 1 & 0 & 0 & 0 & 0 \\ 1 & 0 & 1 & 0 & 0 & 0 \\ 0 & 1 & 0 & 1 & 0 & 0 \\ 0 & 0 & 1 & 1 & 1 & 0 \\ 0 & 0 & 0 & 1 & 1 & 1 \\ 0 & 0 & 0 & 0 & 1 & 0 \end{bmatrix}$$

In a maximal length CA with rule 90 and 150, some special type of configurations are always part of the same cycle:

Definition 2. *A p-configuration is defined as a specific configuration where only one cell is set to 1, and all other cells are set to 0. It is denoted as p^i, where 'i' ranges from 0 to $n - 1$, signifying that the ith cell from the right is the one set to 1.*

For example, in a 6-cell CA, the p-configurations would be 000001, 000010, 000100, 001000, 010000 and 100000. In our approach, we employ characteristics matrix and p-configurations to enable parallelization.

2.2   The Tempering Function

To enhance the randomness of our PRNG and to bolster its performance in empirical tests, we apply a tempering function to the output generated from the CA. This function is borrowed from the 64-bit version of the Mersenne Twister MT-19937 [6]:

$$y = (num \oplus ((num >> u) \ \& \ d)) \tag{1}$$
$$y = (y \oplus ((y << s) \ \& \ b)) \tag{2}$$
$$y = (y \oplus ((y << t) \ \& \ c)) \tag{3}$$
$$res = (y \oplus (y >> l)) \tag{4}$$

where num the is present configuration of the CA, $u = 29$, $s = 17$, $t = 37$, $l = 43$, $b = 0x71d67fffeda60000$, $c = 0xfff7eee000000000$, $d = 0x5555555555555555$ and res is the output of the PRNG after tempering.

2.3   Testing

We subject our proposed PRNG thorough empirical examination using the stringent Dieharder and NIST test suites. A PRNG passes a test suite if all its tests are passed. However, almost all of the existing PRNGs, including the Mersenne Twister, fail to achieve this [1]. Here is a concise overview of the empirical testing procedures:

Dieharder: Dieharder [10] is an advanced test suite designed as an enhancement to George Marsaglia's Diehard test suite for evaluating randomness. It encompasses a comprehensive battery of 114 empirical tests crafted to scrutinize the quality of random sequences. These tests cover a range of assessments, including the Birthdays Test, the BitStream test, and the 3D Sphere test. Dieharder operates by taking a binary file of around 1 GB containing the numbers generated by the PRNG as its input for assessment. Here, we consider 1.5 GB as the size of the .*bin* file.

NIST: Here, a total of fifteen statistical tests were developed, implemented and evaluated that contains a total of 188 subtests [11]. These tests includes Frequency (Monobits) Test, the Binary Matrix Rank Test, Serial Test, Cumulative Sum (Cusum) Test, etc. This test suite accepts an input file containing numbers generated by the PRNG, which can be in either binary or ASCII representation. However, here we use the same binary file of 1.5 GB that is used in Dieharder.

3   Cellular Automata as Multiple Stream Parallel PRNG

Each configuration of the CA corresponds to an internal state of the PRNG and an output is generated from the internal state. So, an internal state evolves from one configuration to the next, following the sequence of rules of the CA. Our chosen CA is a maximal length CA of 64 bits constructed following the logic of Ref. [9] – if we apply rule 150 to the third and fifth cells, and rule 90 to all other cells, the cycle length of the CA is maximal ($2^{64} - 1$). As each configuration directly corresponds to an internal state and an output of the PRNG, for this CA, the period of the PRNG is also $2^{64} - 1$, that is, maximal.

The conventional approach of implementing CA on computer is using an n-length array to represent the n cells and apply rules individually over the array elements. This approach involves a significant computational overhead totally destroying the inherent parallelism of cellular automata. So, to make a fast implementation of PRNGs, this computational overhead needs to be minimized. For our case, conventional implementation means working with a 64-length array where each array indices stores either 0 or 1, that is, a bit. Therefore, instead of considering each bit individually, if we can treat the whole 64-bit configuration as a single 64-bit number and apply bitwise operation of the number to implement the rules of the CA, it will create a tremendous speedup on the computational time and preserve the inherent parallelism of CAs. For this, we take a similar approach like Ref. [7]. Let us explain the implementation process. As the CA has three neighborhood dependency, for applying bitwise operations, first we need to align the neighbors on the same position. In terms of bitwise implementation, left neighboring cell $i - 1$ for any cell i can be found in the same bit position i if we just left shift the configuration (n-bit number) by one bit (operation ($pc << 1$) in the pseudo-code). Similarly, the right neighbor can be put into the same position by just right shifting the configuration by one bit (operation ($pc >> 1$)). Here, as null boundary is considered, circular shift is not required. Now, application of rule 90 to cell i means XOR of the states of cell $i - 1$ and cell $i + 1$. Whereas, rule 150 effectively does XOR over the states of cell $i - 1$, i and $i + 1$. Once we align all the neighbors on the same position by using shifted numbers, we can directly XOR these numbers to apply either rule 90 (left shifted number $\oplus$ right shifted number) or rule 150 (XOR of all three numbers) to all cells in parallel. However, here, as some of the cells use a different rule, we need some more tricks to incorporate this non-uniformity. For example, as the third cell uses rule 150, we need to do an additional XOR operation with the original third bit of the present configuration for only that cell. This can be done if we extract the third bit value as it is, keeping rest of the bits as zero, as XOR with zero does not create any effect. To do this, we AND the present configuration with a number having only 1 at the third bit with the rest of the bits as zero (operation $thirdBit = pc \, \& \, 0x2000000000000000$ in the pseudocode; here the value $0x2000000000000000$ is ($1 << (64 - 3)$) in HEX). Similarly, we can do the same for the fifth bit and finally XOR all these new numbers together to get the updated configuration. Below is the pseudo-code for the bitwise implementation of this CA using the combination of rule 90 and 150:

$$thirdBit = pc \, \& \, 0x2000000000000000 \tag{5}$$

$$fifthBit = pc \, \& \, 0x0800000000000000 \tag{6}$$

$$nc = (pc << 1) \oplus (pc >> 1) \oplus thirdBit \oplus fifthBit \tag{7}$$

Here, pc is the present configuration and nc is the next configuration of the CA. There are two temporary values – $thirdBit$ and $fifthBit$; $thirdBit$ (respectively, $fifthBit$) is a modified number where only the third (respectively, fifth) bit is same as the third (respectively, fifth) cell value of pc.

As a sample example for a 12-bit CA, let 001010101011 be a configuration of the CA where third and fifth cells use rule 150 and other cells use rule 90. Then the temporary number for left neighbor is 010101010110 generated by ($pc <<$ 1) and right neighbor is 000101010101 generated by ($pc >> 1$). The $thirdBit$ is 001000000000 and $fifthBit$ is 000010000000. By the above logic, the next configuration is 010101010110 $\oplus$ 000101010101 $\oplus$ 001000000000 $\oplus$ 000010000000 = 011010000011 which can also be derived by applying the rules cell by cell.

3.1   Division of Cycle Length into Multi-threads

To enable the parallelization of this PRNG, it is essential to divide the cycle length into multiple threads, with each thread responsible for computing a portion of the cycle of numbers. However, the key question is how to segment the cycle effectively to ensure parallel execution, determine the seeds for each thread, and decide the number of threads.

Approach 1: In this approach, we can run a maximum of n threads in parallel, with each thread commencing from a distinct p-configuration as the seed. These threads continue their execution until they encounter another p-configuration, at which point the process is terminated, and the thread ceases its operation. However, this approach had a notable drawback: the positions of the p-configurations in the cycle are not uniformly distributed. So, the segment lengths are not equal, leading to situations where some threads are terminated quickly, while others are running for considerably longer duration.

Approach 2: To ensure equal segment length, a systematic approach involves selecting seeds, $s_1, s_2, \cdots, s_n$, which are equally spaced within the cycle (see Fig. 1). The aim is to set thread lengths as v so that they match the distance between any two of these seeds in the sequence. The process of finding this v^{th} configuration from the initial configuration is known as *jumping ahead* [7].

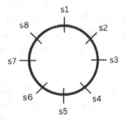

Fig. 1. Selecting equally spaced seed from the cycle of maximal length CA

In this jumping ahead method, a seed is chosen initially, and then we leap forward by v steps within the cycle to determine k equally spaced seeds. The idea here is to accommodate the user's preference for running k threads in parallel. For a maximal length cellular automaton, the relationship between this jump size as well as equal segment size v, and k, the number of threads, is as follows:

$$v = (2^n - 1)/k \tag{8}$$

To execute this process efficiently, we leverage the characteristic matrix of the CA. By multiplying the tth configuration represented as a bitwise vector (of order $1 \times n$) with the characteristic matrix T (of order $n \times n$), we obtain a vector of order $1 \times n$ signifying the CA's configuration at the next time step, $t+1$. Likewise, if we multiply this vector with the square of the characteristic matrix, we can acquire the configuration at time $t + 2$, and so on. In essence, by multiplying the current configuration vector with a matrix T^k, one can obtain the CA's next configuration at time $t + k$. For the jumping ahead method, we pre-compute the values of $T^v, T^{2v}, \cdots, T^{(k-1)v}$ and multiply them with the vector representation of the seed (initial configuration of the CA). This results in the generation of all k seeds to initialize the k threads where value of k depends on user's choice and configuration supported by the used machine.

However, for the current 64-bit multi-core machines, maximum number of supported threads are always power of two. So, if we take the number of parallel streams or threads (k) as power of 2, especially 2, 4, 8, 16, 32, 64, we can precompute the jump size v for each of these options and the values of T^v, $T^{2v}, \cdots, T^{(k-1)v}$ for that v. Then, user can be given as choice these options as number of threads to choose from. This approach significantly streamlines the distribution of computational costs compared to the previous methods. As this pre-computation is one time only and not part of the random number generation, significant speed-up can be achieved even when number of parallel streams is only 2. However, it is important to note that, this PRNG using only linear rules (with or without equal length threads) is not cryptographically secure.

3.2 Algorithmic Details

The user can enter the number of threads, k and one seed value, s. Subsequently, the PRNG can employ the jumping-ahead mechanism to compute the other $k-1$ seeds and initialize the k threads for the concurrent generation of k random numbers at any given time. Otherwise, it can randomly select k number of p-configurations to be assigned to the k threads as seeds.

Approach 1. Using p-Configuration as Seed for Each Thread: In this approach, we generate random numbers using k threads having their initial configuration as one of the p-configurations. Each thread runs independently to produce random numbers until it reaches any of the other p-configurations chosen as seed. The details of the algorithm is shown in Algorithm 1.

Here, for each thread, $LHCA64(presentConfiguration)$ is the function to generate the next configuration of the CA from the present configuration taken as argument and return the next configuration. The next configuration is then passed through the $temper()$ function to generate the output of the PRNG. However, this does not change the internal state of the PRNG which is same as the current configuration of the CA. So, the next configuration is fed again as input to $LHCA64()$ and the number generation process continues untill any of the p-configurations taken as seeds are encountered. In that case, as we know all p-configurations are part of the same cycle for this CA, reaching same p-configuration means the number will be repeated after that. So, that thread is stopped.

Approach 2. Dividing Whole Cycle into k Equal Segments: In this approach, based on the value of k, $k-1$ seeds are calculated using jumping ahead property using the user given seed and T matrix. Each thread run independently to produce random numbers until it reaches some other seed or completes the

---

Algorithm 1. Threading with p-configuration

---

setOfSeeds ← {}
outputFile ← output.bin
function GENERATERANDOMNUMBERS(threadAddress, seed)
    $randomNumber$ ← $seed$
    repeat
        $randomNumber$ ← LHCA64($randomNumber$)
        //$LHCA$64 uses Equation 5 to generate next configuration
        $temperedNumber$ ← temper($randomNumber$)
        //$temper$ uses Equation 1 for tempering
        outputFile.write($temperedNumber$) // Critical section for threads
    until $randomNumber \notin$ setOfSeeds
end function

function SEEDINITIALIZATION(K)
    for $i \in \{0, 1, 2, \ldots, k-1\}$ **do**
        $setOfSeeds$.append(2^i) //p-configurations as seed
    end for
    $threads$ ← $\{t_0, t_1, t_2, \ldots, t_{k-1}\}$
    for $i \in \{0, 1, 2, \ldots, k-1\}$ **do**
        $randomNumber$ ← GenerateRandomNumbers($\&t_i$, $setOfSeeds[i]$)
    end for
    for $i \in \{0, 1, 2, \ldots, k-1\}$ **do**
        joinThread(t_i) //Wait for thread to complete
    end for
end function

---

segment length v. Algorithm 2 gives the details of this implementation (here, we use the implementation of function $generateRandomNumbers()$ similar to Algorithm 1).

Here also, the same $LHCA64()$ and $temper()$ functions are used. The only difference is in the initialization of seeds. Here, based on user's choice of the number of threads (k), jumping ahead mechanism (see Approach 2 in Page 7 for details) is utilized to calculate the other $k-1$ seeds from a given seed. However, as mentioned, we can totally avoid this jumping ahead mechanism that includes costly matrix multiplication in the run-time by pre-computing the matrices for some predefined number of threads as power of 2. Then, user can be given only those number of threads as options to choose from.

3.3 Implementation

The practical details of our PRNG implementation are as follows:

      Internal State: Each thread's internal state is represented by the n-bit cellular automaton (CA) configuration.

Output Generation: At each time step, the PRNG produces k 64-bit random numbers in a k-threaded implementation.

Algorithm 2. Algorithm for Equally Spaced Segments

---
setOfSeeds ← {}
outputFile ← output.bin
function GENERATERANDOMNUMBERS(threadAddress, seed)
    randomNumber ← seed
    repeat
        $randomNumber$ ← LHCA64($randomNumber$)
        //$LHCA64$ uses Equation 5 to generate next configuration
        $temperedNumber$ ← temper($randomNumber$)
        //$temper$ uses Equation 1 for tempering
        outputFile.write($temperedNumber$) // Critical section for threads
    until randomNumber $\notin$ setOfSeeds //Or, segment size v is exhausted
end function

function JUMPAHEAD(seed, T, pow)
    return seed $\times T^{\text{pow}}$
end function

function SEEDINITIALIZATION(SEED,K)
    randomNumber ← seed
    $v \leftarrow (2^n - 1)/k$
    for $i \in \{1, 2, \ldots, k\}$ **do**
        $s \leftarrow$ jumpAhead(seed, T, v^i) // Generate other equally spaced seeds
        setOfSeeds.append(s)
    end for
    threads ← $\{t_0, t_1, t_2, \ldots, t_{n-1}\}$
    for $i \in \{1, 2, \ldots, k\}$ **do**
        randomNumber ← generateRandomNumbers($\&t_i$, setOfSeeds[i])
    end for
    for $i \in \{1, 2, \ldots, k\}$ **do**
        joinThread(t_i) //Wait for completion of thread
    end for
end function

Transition Function: To determine the new internal state of the generator, the chosen set of rules is used as the transition function.

Tempering: Following the generation of a new internal state of the PRNG (configuration of the CA), a series of transformations, named tempering, is applied to yield the random number at each time step.

Multi-Threading: To enable multi-threading and initialize various threads, two methods are employed to generate seeds for each thread:

1. A set of k threads, each commencing with a unique p-configuration as the initial state.
2. Utilizing the jump-ahead algorithm to generate k seeds for the threads. This can also be customized for allowing only a specific number of threads.

Comparison: The outcomes of this implementation are tested by benchmarks against existing implementations to assess its performance and the quality of generated random numbers using the Dieharder and NIST test suites.

3.4   Result

We compare our PRNG with MT19937 and SFMT19937 [12], the two most celebrated versions of Mersenne Twisters, both generating 64 bit numbers as output. As these two are single stream PRNG, we made our PRNG to run on only a single thread. All these PRNGs are tested on Dieharder and NIST over a file size of 1.5 GB generated using the default seed value for MT-19937 and SFMT19937 that is, 19650218. Table 1 shows the result of Dieharder test for the first approach where we initialize only one thread with the seed as some p-configuration with temper function and without using the temper function. Similarly, Table 2 shows the result for this approach over the NIST Test Suite.

Table 1. Result of the Dieharder Test Suite for the 1st Approach

Total Test Cases:	Passed	Failed	Weak
114 (With Temper)	79	31	4
114 (Without Temper)	78	31	5

Table 2. Result of the NIST Test Suite for the 1st Approach

Total Test Cases:	Passed	Failed
188 (With Temper)	188	0
188 (Without Temper)	144	44

For the second approach, where a seed is to be provided by the user, we choose the same default seed of MT19937 for ease of comparison. Here, Table 3 and Table 4 show the result for approach 2 under Dieharder and NIST respectively where we initialize a single thread with 19650218 as seed. As, this seed may not be the best seed for our PRNG, testing over other seed can give even a better result for our proposed PRNG.

Table 3. Result of the Dieharder Test Suite for the 2nd Approach

Total Test Cases:	Passed	Failed	Weak
114 (With Temper)	107	2	5
114 (Without Temper)	78	32	4

Table 4. Result of the NIST Test Suite for the 2nd Approach with seed 19650218

Total Test Cases	Passed	Failed
188 (With Temper)	185	3
188 (Without Temper)	148	40

We can clearly see that, our PRNG generated by the approach 2 (Algorithm 2) is a good PRNG with respect to both Dieharder and NIST. Therefore, approach 2 with tempering is our choice of proposed PRNG for 64-bit computing. We now compare this PRNG's performance with Mersenne Twister's family, namely the MT-19937 and SFMT-19937, both of 64-bit output versions. Table 5 and Table 6 depict the results for this comparison. We can see that, our proposed PRNG performs at par with both these PRNGs of MT family even for a single thread implementation.

Table 5. Result of the NIST Test Suite for the Mersenne Twister and SFMT 64 bit versions along with our proposed PRNG for seed = 19650218

PRNG	Total Test Cases	Passed	Failed
MT-19937	188	188	0
SFMT-19937	188	188	0
Proposed PRNG	**188**	**185**	**3**

Table 6. Result of the Dieharder Test Suite for the Mersenne Twister and SFMT 64 bit versions along with the proposed PRNG

PRNG	Total Test Cases	Passed	Weak	Failed
MT-19937	114	112	2	0
SFMT-19937	114	111	3	0
Proposed PRNG	**114**	**107**	**5**	**2**

We now do a comparison over the execution time of the PRNGs. We run all these three PRNGs over Apple M2 Max Laptop with 64 GB RAM and macOS sonoma version 14.0 and generate some sequence of numbers. Our PRNG is made to run on a single thread for comparison. Table 7 shows the result of speed test for our proposed PRNG (LHCA64) with MT19937-64 and SFMT19937-64. We can see that, even for a single thread, our PRNG is at least as fast as the MT and much faster than SFMT for large numbers. This shows the effectiveness of our PRNG has a generator having good randomness quality, lighweightness and speed with capability to generate multiple stream of numbers parallelly.

Table 7. Speed Test of Proposed PRNG with MT and SFMT (64 bit)

Count of Numbers Generated	Execution Time in microseconds		
	LHCA64	MT19937-64	SFMT19937-64
1	0	5	17
10	0	6	17
100	1	8	18
1000	6	45	111
5000	86	78	188
10000	215	442	875
50000	355	1,998	1,641
100000	1,418	1,546	4,583
500000	3,474	8,568	16,370
1000000	6,577	13,977	23,998
5000000	35,888	41,521	77,019
10000000	69,023	61,427	1,28,823
50000000	2,22,416	2,16,589	5,30,850
100000000	4,13,935	4,09,882	10,12,746
500000000	19,54,213	19,20,259	49,94,589
1000000000	38,75,381	37,85,694	99,55,426

4   Conclusion and Future Work

We have designed a 64-bit multiple stream parallel random number generator that harnesses user-supplied seed values and the number of threads to concurrently generate 64-bit random numbers. This parallelization leverages the *jumping ahead* characteristic of CAs, which allows us to create distinct seed values for each thread segmenting the period into equal distributions. As a result, our generator offers substantial speedup when compared to other PRNGs. Moreover, it boasts an impressive periodicity of $2^{64} - 1$, ensuring an extensive cycle of generated numbers. Notably, our PRNG also excels in producing high-quality randomness. It has demonstrated its robustness by successfully passing the majority of the rigorous tests within the NIST and Dieharder test suites, confirming its reliability and suitability for a wide range of applications. Because of its lightweight implementation, it is well-suited for applications in current multi-core CPUs. However, the task to evaluate our PRNGs using the BigCrush empirical test suite and theoretical testing remains to be done. Furthermore, work may be directed to improve randomness quality without using the tempering function.

Acknowledgment. This work is partially supported by Start-up Research Grant (File number: SRG/2022/002098), SERB, Govt. of India. The authors are grateful to Prof. Sukanta Das for his valuable comments and guidance throughout the Summer School and even after, for completing this work. A special thanks to Subrata Paul for helping with the empirical tests.

References

1. Bhattacharjee, K., Das, S.: A search for good pseudo-random number generators: survey and empirical studies. Comput. Sci. Rev. **45**, 100471 (2022)
2. Matsumoto, M., Nishimura, T.: Mersenne twister: a 623- dimensionally equidistributed uniform pseudo-random number generator. ACM Trans. Model. Comput. Simul. (TOMACS) **8**(1), 3–30 (1998)
3. Wolfram, S.: Origins of randomness in physical systems. Phys. Rev. Lett. **55**(5), 449 (1985)
4. L'Ecuyer, P., Nadeau-Chamard, O., Chen, Y.-F., Lebar, J.: Multiple streams with recurrence-based, counter-based, and splittable random number generators. In: 2021 Winter Simulation Conference (WSC), pp. 1–16. IEEE (2021)
5. Salmon, J.K., Moraes, M.A., Dror, R.O., Shaw, D.E.: Parallel random numbers: as easy as 1, 2, 3. In: Proceedings of 2011 International Conference for High Performance Computing, Networking, Storage and Analysis, pp. 1–12 (2011)
6. More, N., Singh, S.K., Verma, N., Bhattacharjee, K.: Cellular automaton-based emulation of the mersenne twister. Complex Syst. **32**(2), 139–169 (2023)
7. Jaleel, H.A., Kaarthik, S., Sathish, S., Bhattacharjee, K.: Multiple-stream parallel pseudo-random number generation with cellular automata. In: Manzoni, L., Mariot, L., Roy Chowdhury, D. (eds.) Cellular Automata and Discrete Complex Systems. AUTOMATA 2023. LNCS, vol. 14152, pp. 90–104. Springer, Cham (2023). https://doi.org/10.1007/978-3-031-42250-8_7
8. Cattell, K.M., Muzio, J.C.: Table of Linear Cellular Automata for Minimal Weight Primitive Polynomials of Degrees Up to 300. University of Victoria, Department of Computer Science (1991)
9. Cattell, K., Zhang, S.: Minimal cost one-dimensional linear hybrid cellular automata of degree through 500. J. Electron. Test. **6**(2), 255–258 (1995)
10. Brown, R.G., Eddelbuettel, D., Bauer, D.: Dieharder. Duke University Physics Department, Durham, NC. www.webhome.phy.duke.edu/~rgb/General/dieharder.php
11. Rukhin, A., et al.: A statistical test suite for random and pseudorandom number generators for cryptographic applications, volume 800–22 (revision 1a). National Institute of Standards and Technology, Technology Administration, U.S. Department of Commerce (2010)
12. Saito, M., Matsumoto, M.: SIMD-oriented fast mersenne twister: a 128-bit pseudorandom number generator. In: Keller, A., Heinrich, S., Niederreiter, H. (eds.) Monte Carlo and Quasi-Monte Carlo Methods 2006, pp. 607–622. Springer, Heidelberg (2006). https://doi.org/10.1007/978-3-540-74496-2_36

A Dual-Image Based Secured Reversible Data Hiding Scheme Exploiting Weighted Matrix and Cellular Automata

Kankana Datta[1]($\boxtimes$), Som Banerjee[2], Biswapati Jana[3], and Mamata Dalui[2]

[1] Haldia Institute of Technology, Purba Midnapore, Haldia, West Bengal, India
dattakankana18@gmail.com
[2] National Institute of Technology, Durgapur, West Bengal, India
sb.21cs1105@phd.nitdgp.ac.in , mdalui.cse@nitdgp.ac.in
[3] Vidyasagar University, Midnapore, West Bengal, India

Abstract. In the era of advanced communication technology, security of confidential data during transmission through public media is an important issue. To curtail the risk of attacks, a secured, reversible, robust data concealing scheme with good visual quality and embedding capacity is essential while improving execution efficacy. In this context, a Cellular Automata based reversible and secured data hiding scheme using weighted matrix has been applied on dual-image which provides a trade-off between embedding capacity and visual quality. Since, the proposed scheme is robust enough against various venomous attacks, this approach is useful for private and public sectors practitioners to protect important multimedia secret information from adversarial cyber attacks.

Keywords: Data hiding · Weighted Matrix · Cellular Automata · Secret Message · Reversibility

1 Introduction

Now-a-days, the internet usage has exacerbated expeditiously due to the aggressive information sharing through social media. Hence, safe and secured transmission of multimedia documents has become a challenging issue for the E-world. While transmitting the secret data, various activities, such as malicious tampering [1], illegal access [2] etc. are observed due to some security loopholes. To get-rid-of these existing security issues, one of the popular technique, called steganography is used. Steganography [3] is the technique of hiding secret information within a innocuous digital cover media (text, image, audio, video etc.) with minimum distortion of its visual quality such that it becomes hard to detect the existence of the secret information. The concealed media is called stego media which is almost identical to the cover media with the hidden information. A robust and efficient steganography scheme always takes care of a secure transmission of secret data which is important for different purposes in real life such as tamper detection [1], medical images [4], digital forensics [5], etc. Though,

M. Dalui et al. (Eds.): ASCAT 2024, CCIS 2021, pp. 123–136, 2024.
https://doi.org/10.1007/978-3-031-56943-2_10

there are many popular steganographic schemes available to prevent the secret message efficiently, but, due to some loopholes, they are unable to meet the current market demand completely. This has motivated us to propose an image steganographic scheme which outperforms the other state-of-the-art schemes in terms of embedding capacity, reversibility, visual quality and robustness against different types of malicious attacks like filtering, modification and compression, etc. The major objectives and motivations of the proposed scheme are given below:-

- **Imperceptibility**:- One of the essential characteristics of any data hiding technique is the imperceptibility. The concealed secret data within cover image should not cause any degradation in visual quality of the image.
- **Reversibility**:-After embedding, the cover media may get damaged and it becomes hard to recover at the receiver's end. Recovery of cover media is important in various human centric applications. In such applications, instead of conventional data hiding, reversible data hiding schemes are preferred.
- **Embedding Capacity**:- High embedding capacity, an essential criteria for any steganographic solution, can be achieved by using dual cover images.
- **Security**:- The secret messages are distributed in both the cover images by embedding technique in such a way that without any one of them, it is impossible to recover the secret message and the original cover image during extraction. So, sharing of information via dual images help to enhance the security.
- **Robustness and Temper Detection**:- A scheme should be robust against compression, modification, filtering, noise, contrast etc. It must be free from unauthorized detection and decoding. It should resist tampering or would resist only up to a certain extent. Temper detection and correction are essential to recover the secret data by authorized user.

Though dual-image based schemes are computationally more expensive than single-image based schemes as they involve processing of two images, but, this additional complexity can provide greater advantages for enhancement of security, higher data hiding capacity and better robustness against attacks in the area of advanced communication technology as compared to other state-of-the-art schemes.

This paper is organized into various sections and subsections which provide the detailed description of the proposed work. The Sect. 1 formally introduces the work and preliminaries of Cellular Automata (CA) are reported in Sect. 2. Section 3 describes the proposed scheme in detail including algorithms and numerical examples of embedding and extraction techniques. Section 4 comprises of the experimental results, analysis and performance comparison of the proposed scheme with other existing schemes. At last, conclusions are drawn in Sect. 5.

2 Preliminaries of Cellular Automata (CA)

Cellular automata (CA) is a distinct form of finite state machine which is denoted as uniform array of identical cells in n-dimensional space. Synchronously, the cells change their states at a specific time instant. Figure 1 (a) represents a periodic boundary CA (PBCA). CA can be applied in different fields such as image encryption [6], quantum computation [7], etc.

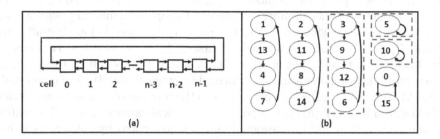

(a)   (b)

Fig. 1. (a) Periodic boundary CA. (b) State transition diagram of a 4-cell uniform $PCBA$ configured with rule 85 (3 unity attractors are highlighted by red dashed boxes). (Color figure online)

A linear array of cells with 3 neighbourhood dependency is a special case, called Elementary cellular automata (ECA), where the state of each cell is denoted by either 0 or 1. The status of a CA during its dynamic evolution exihibits a cyclic pattern which is called attractor. Some ECA rules form attractors, called unity attractors [6,8], which exhibit the property that applying XOR operation to all configurations (states) of a cycle (attractor) yields zero. This is represented as $(\oplus_{s=1}^{y} p^s) = 0$. Here, p^s is the cell state of an n-cell CA at time s, where, $1 \leq s \leq y$ and y is the number of states of one of such attractors. The bitwise XOR between the states (binary representation) in each attractor results into zero. Figure 1 (b) represents the state transition diagram of a 4-cell Elementary Cellular Automata (ECA) configured with rule 85 in periodic boundary condition. In this figure, three unity attractors are indicated by red dashed boxes. By utilizing this special property of unity attractor, we can derive the following

$\alpha \oplus (\oplus_{s=1}^{t} p^s) = \alpha'$ and $\alpha' \oplus (\oplus_{s=t+1}^{y} p^s) = \alpha$ where $1 \leq t \leq y$ and α and α' both are n-bit binary data. Here, the XOR operation is performed between α with some of the states within an attractor basin (from s to t) and output is α'. Then, α' is XORed with the remaining states (from $t+1$ to y) of the same attractor basin which generates the original value α. This phenomenon proves that $(\oplus_{s=1}^{t} p^s)$ and $(\oplus_{s=t+1}^{y} p^s)$ generate the same value and this property of unity attractor of rule 85 has been utilized as a basis for key generation in the proposed scheme.

3   The Proposed Scheme

This section presents the proposed scheme in detail. The embedding and extraction techniques are elaborated in the following subsections.

3.1   Embedding Technique

A $(m \times n)$ cover image I is divided into (3×3) non-overlapping, equal square image matrices of pixels called bitmaps. Each bitmap is represented by B_i where i represents the sequence number of bitmaps. Complete embedding algorithm of the proposed scheme with a numerical example is discussed below. Initially, first bitmap $B_{i=1}$ is selected and a weighted matrix $W_{i=1}$ of size (3×3) is created by taking the unit digit of each cell of B_i. A summation of cell by cell weighted matrix multiplication operation is performed between B_i and W_i and computed ξ by $\Sigma(B_i(x,y) \times W_i(x,y))\%2^r$, where $r = 4$, is the number of bits to be embedded. This is shown in Fig. 2 (a) and as per this numerical example $\xi = 7$. Thereafter, the value $B_i(2,2) = 149$ is considered and α is generated by converting it into 8-bit binary as shown in Fig. 2 (b). Here, the value of α is 10010101. To generate a key, an unity attractor of 8-cell $PBCA$ configured with rule 85 is used. The CA is run with an initial seed of 01100110.

We have chosen $key = \delta = \alpha \oplus (\oplus_{s=1}^{t}p^s)$, where $1 <= t <= 4$ and p^s is the configuration of CA at instant s. Using the property of unity attractor as discussed in Sect. 1, it is possible to generate and retrieve the same key value during embedding and extraction times respectively. At the time of embedding, XORing is performed on t number of states, say, 1st to t^{th} state, of the unity attractor of the 8-cell CA $< 85, 85, \ldots, 85 >$ while for extraction, the XORing is performed on the remaining states, say $(t+1)^{th}$ to y^{th} state, of the same attractor. Now, as per the property of unity attractor, both XOR operations should produce same results, which helps to retrieve the key value during extraction time. Here, the value of t is chosen as either 2 or 3, based on the odd/even value of α. Since, in this example 149 is an odd number, 2 is chosen as a value of t. The key value is represented by δ_n, where n is the index of each bits from 1,2,3,...,8. In Fig. 2 (b), the value of δ is computed as 11000000. 32 bits are selected from secret message M sequentially and divide them into 8 chunks where each holds 4 bits and generate σ_q by converting each chunks into respective decimal values, where $q \leftarrow 1, 2, 3\ldots, 8$. This is depicted in Fig. 2 (c) and the values from σ_1 to σ_8 are 12, 15, 10, 3, 7, 1, 9 and 14. If, q is 1, then λ_q is computed by $(\sigma_q - \xi)$, otherwise, by $(\sigma_q - \sigma_{q-1})$. For each σ_q, there is a value for λ_q. As per Fig. 2 (c), the eight values from λ_1 to λ_8 are 5, 3, -5, -7, 4, -6, 8 and 5. This λ_q is used to modify the value of actual pixels at the time of embedding. Thereafter, two copies of bitmap B_i, namely AB_i and EB_i are created as shown in Fig. 2 (c). The secret data are embedded into the dual bitmaps AB_i and EB_i by using the operations defined in Algorithm 1 (from line number 21 - 47) on each cell, except the cells $AB_i(2,2)$ and $EB_i(2,2)$, keeping in mind the conditions $(MP_i(x,y) \geq 0)$ and $(MP_i(x,y) \leq 255)$ as underflow and overflow respectively. $AP_i(x,y)$ and $MP_i(x,y)$ represent Actual Pixel and Modified Pixel respectively. $MP_i(x,y)$ is

---

Algorithm 1: Secret data embedding algorithm for a cover bitmap B_i of the proposed scheme.

Input : A (3×3) cover bitmap B_i, Secret message M.
The following algorithm is used for a bitmap B_i and iterated
for remaining bitmaps.

Output: Two stego bitmaps namely $ActualBitmap$ (AB_i) and $ExtendedBitmap$ (EB_i) each of size (3×3).

```
 1  Function Embedding(B_i, M):
 2  |    Create weighted matrix (W_i) by unit digit of each cell value of B_i.
 3  |    ξ = Σ(B_i(x, y) × W_i(x, y))%2^r [r = number of bit to be embedded = 4].
 4  |    Generate α by converting B_i(2, 2) into 8 bits binary numbers.
 5  |    if (α % 2 == 0) then
 6  |    |    set t ← 3.
 7  |    else
 8  |    |    set t ← 2.
 9  |    end
10  |    An unity attractor of 8-cell PBCA configured with rule 85 is used to generate the key and fix the
    |    initial seed.
11  |    key = δ = α ⊕ (⊕_{s=1}^{t} p^s) where 1 <= t <= 4 and p^s is the configuration of CA at instant s.
12  |    Select 32 bits from secret message M sequentially and divide them into 8 chunks,each holds 4 bits
    |    and generate σ_q by converting each chunks into respective decimal values, where, q ← 1, 2, 3, ....8.
13  |    for (q = 1; q <= 8; q + +) do
14  |    |    if (q == 1) then
15  |    |    |    set λ_q ← σ_q − ξ.
16  |    |    else
17  |    |    |    set λ_q ← σ_q − σ_{q−1}.
18  |    |    end
19  |    end
20  |    Assign the index numbers on key δ_n where n represents the index of each bits from 1,2,3,...,8.
21  |    set ro ← 3; set cl ← 3; set q ← 1; set n ← 1;
22  |    for (x = 1; x <= ro; x + +) do
23  |    |    for (y = 1; y <= cl; y + +) do
24  |    |    |    if ((x == 2) && (y == 2)) then
25  |    |    |    |    continue;
26  |    |    |    else
27  |    |    |    |    if (δ_n == 1) then
28  |    |    |    |    |    MP_i(x, y) ← AP_i(x, y) + λ_q;
29  |    |    |    |    |    if ((MP_i(x, y) ≥ 0) || (MP_i(x, y) ≤ 255)) then
30  |    |    |    |    |    |    AB_i(x, y) ← AP_i(x, y); EB_i(x, y) ← MP_i(x, y);
31  |    |    |    |    |    |    q ← q + 1; n ← n + 1;
32  |    |    |    |    |    else
33  |    |    |    |    |    |    Go to the next bitmap B_{i+1}.
34  |    |    |    |    |    end
35  |    |    |    |    else
36  |    |    |    |    |    MP_i(x, y) ← AP_i(x, y) + λ_q;
37  |    |    |    |    |    if ((MP_i(x, y) ≥ 0) || (MP_i(x, y) ≤ 255))) then
38  |    |    |    |    |    |    AB_i(x, y) ← MP_i(x, y); EB_i(x, y) ← AP_i(x, y);
39  |    |    |    |    |    |    q ← q + 1; n ← n + 1;
40  |    |    |    |    |    else
41  |    |    |    |    |    |    Go to the next bitmap B_{i+1}.
42  |    |    |    |    |    end
43  |    |    |    |    end
44  |    |    |    end
45  |    |    end
46  |    end
47  |    temp ← EB_i(2, 2); EB_i(2, 2) ← EB_i(2, 3); EB_i(2, 3) ← temp.
```

---

generated by summation of $AP_i(x, y)$ and λ_q, where q=1,2,3...,8. In Fig. 2 (c), after modifying the pixel values of the bitmaps, swapping operation is performed between $EB_i(2, 2)$ and $EB_i(2, 3)$ that is between 177 and 149. Thus, dual-stego bitmaps AB_i and EB_i are produced at the end of the embedding procedure. The step-by-step embedding technique is presented in Algorithm 1. This embedding technique is applied on all the bitmaps B_i sequentially. Finally, all the stego bitmaps AB_i and EB_i are concatenated individually including all unmodified pixels of the cover image and dual-stego images SI' and SI'' are produced for sending to the receiver.

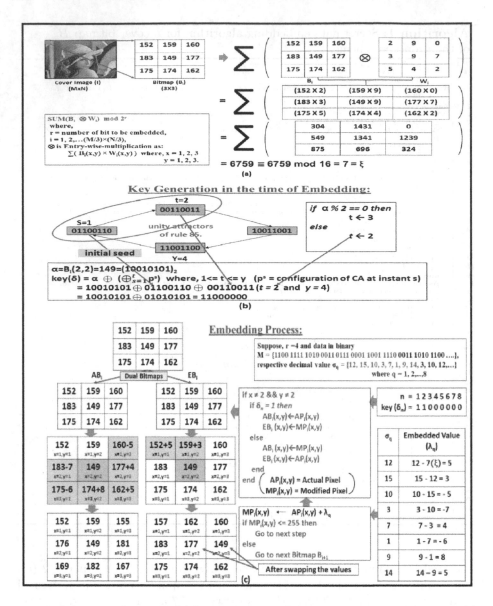

Fig. 2. Schematic diagram of (a) Summation of weighted matrix to generate ξ, (b) Key generation using PBCA at the time of embedding, (c) Embedding technique for i^{th} bitmap of the proposed scheme.

3.2 Extraction Technique

At the end of the extraction process, the original cover image I and embedded message M both are retrieved successfully from dual-stego images SI' and SI''. This proves the reversibility property of the suggested scheme as achieved by

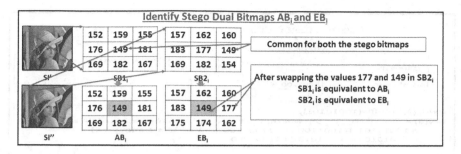

Fig. 3. Schematic diagram of bitmaps identification technique from bitmaps $SB1_i$ and $SB2_i$ of the proposed scheme.

employing CA. The complete extraction technique of the proposed scheme with a numerical example is discussed below. In the extraction phase, the dual-stego images SI' and SI'' are divided into (3×3) non-overlapping, equal bitmaps which are represented by $SB1_i$ and $SB2_i$ respectively where i represents the index of bitmap.

Initially, each pair of cell values of bitmaps $SB1_i$ and $SB2_i$ are compared. If the value of every pair of pixels are same then, no extraction technique is applied on bitmaps $SB1_i$ and $SB2_i$ and moved to the next bitmaps $SB1_{i+1}$ and $SB2_{i+1}$ respectively. Else, following extraction operations are performed on $SB1_i$ and $SB2_i$. To identify the bitmaps AB_i and EB_i, a comparison is made between $SB1_i(2, 2)$ and $SB2_i(2, 3)$. If, they provide same value then, bitmap $SB1_i$ represents AB_i and $SB2_i$ represents EB_i. Otherwise, $SB1_i$ represents EB_i and $SB2_i$ represents AB_i. This is depicted in Fig. 3. As per this figure, as $SB1_i(2, 2)$ and $SB2_i(2, 3)$ both are 149, so bitmap $SB1_i$ represents AB_i and $SB2_i$ represents EB_i. Thereafter, a swapping operation is performed between $EB_i(2, 2)$ and $EB_i(2, 3)$. In Fig. 4(a), $\alpha = 10010101$ is generated by converting the value of $B_i(2, 2) = 149$ into 8-bit binary. To retrieve the key, the property of an unity attractor of a 8-cell $PBCA$ configured with rule 85 is used. The CA is run with an initial seed of 01100110.

Here, $key = \delta = \alpha \oplus (\oplus_{s=t+1}^{y} p^s)$, where $1 <= t <= y$ and p^s is the configuration of CA at instant s. The remaining states of the unity attractor of CA rule 85, which are not used during the embedding time, take part to retrieve the key value during extraction time. Here, the value of t is chosen as either 2 or 3, based on the odd/even value of α. As per this example, t is 2, since $\alpha = 149$ is an odd number. The key value is represented by δ_n, where n is the index of each bits from 1,2,3,...,8. To segregate the actual pixels from modified pixels, we use the operations defined in Algorithm 2 (from line number 18–33) for each cell of AB_i and EB_i except the cells $AB_i(2, 2)$ and $EB_i(2, 2)$. As a consequence, bitmaps AP_i and MP_i are generated which is depicted in Fig. 4 (b). The bitmaps AP_i represents original cover bitmap B_i and MP_i represents modified bitmap with updated pixels. Eight λ_q values from λ_1 to λ_8 are shown in Fig. 4 (b). They are 5, 3, -5, -7, 4, -6, 8 and 5. In Fig. 4 (c), a weighted matrix W_i is created

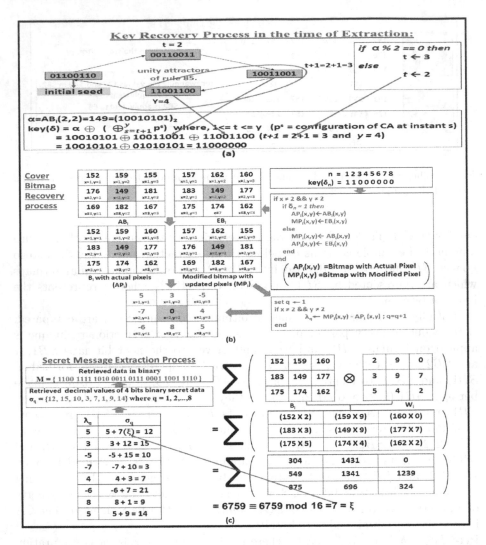

Fig. 4. Schematic diagram of (a) Key recovery technique at the time of extraction using PBCA,(b) Cover Bitmap recovery technique, (c) Secret Message extraction technique using summation of weighted matrix by generating ξ for i^{th} bitmap of the proposed scheme.

by unit digit of each cell of bitmap B_i. A summation of cell by cell weighted matrix multiplication operation is performed between B_i and W_i and generate ξ using $\Sigma(B_i(x,y) \times W_i(x,y))\%2^r$, where $r = 4$ number of bit to be embedded. As per numerical example, the value of ξ is 7. Thereafter, the value of σ_q is computed by $(\lambda_q + \xi)$, while q is 1. For remaining seven values of q, σ_q is $(\lambda_q + \sigma_{q-1})$. According to Fig. 4 (c), the eight decimal values from σ_1 to σ_8 are 12, 15, 10, 3, 7, 1, 9 and 14. These eight decimal values, containing 4 bits binary

---

Algorithm 2: Original cover image and secret data extraction algorithm from the dual-stego bitmaps $SB1_i$ and $SB2_i$ of the proposed scheme.

Input : Dual-stego bitmaps namely $(SB1_i)$ and $(SB2_i)$ each of size (3×3).
            The following steps are repeated for all the stego bitmaps $SB1_i$ and $SB2_i$
            until or unless all secret bits are recovered successfully.
Output: 32 secret data bits and a (3×3) cover bitmap B_i.

```
1   Algorithm Extraction(SB1_i, SB2_i):
2       If, each pair of cell values of bitmaps SB1_i and SB2_i are same, then, the next bitmaps SB1_{i+1}
        and SB2_{i+1} are processed. Otherwise, following extraction operations are performed.
3       if (SB1_i(2,2) == SB2_i(2,3)) then
4       |   SB1_i represents AB_i; SB2_i represents EB_i.
5       else
6       |   SB1_i represents EB_i; SB2_i represents AB_i.
7       end
8       temp ← EB_i(2,2); EB_i(2,2) ← EB_i(2,3); EB_i(2,3) ← temp.
9       Generate α by converting B_i(2,2) into 8 bits binary numbers.
10      if (α % 2 == 0) then
11      |   set t ← 3.
12      else
13      |   set t ← 2.
14      end
15      An unity attractor of 8-cell PBCA configured with rule 85 is used to retrieve the key and fix the
        initial seed.
16      key = δ = α ⊕ (⊕_{s=t+1}^{y} p^s), where 1 <= t <= y and p^s is the configuration of CA at instant s.
17      Assign the index numbers on key δ_n where n represents the index of each bits from 1,2,3,...,8.
18      set ro ← 3; set cl ← 3; set q ← 1; set n ← 1;
19      for (x = 1; x <= ro; x + +) do
20      |   for (y = 1; y <= cl; y + +) do
21      |   |   if ((x == 2) && (y == 2)) then
22      |   |   |   continue;
23      |   |   else
24      |   |   |   if (δ_n == 1) then
25      |   |   |   |   AP_i(x,y) ← AB_i(x,y); MP_i(x,y) ← EB_i(x,y)
26      |   |   |   |   λ_q ← MP_i(x,y) − AP_i(x,y); q ← q + 1; n ← n + 1;
27      |   |   |   else
28      |   |   |   |   MP_i(x,y) ← AB_i(x,y); AP_i(x,y) ← EB_i(x,y)
29      |   |   |   |   λ_q ← MP_i(x,y) − AP_i(x,y); q ← q + 1; n ← n + 1;
30      |   |   |   end
31      |   |   end
32      |   end
33      end
34      AP_i represents recovery cover image B_i.
35      Create weighted matrix (W_i) by unit digit of each cell value of B_i.
36      ξ = Σ(B_i(x,y) × W_i(x,y))%2^r  [r = number of bit to be embedded = 4].
37      for (q = 1; q <= 8; q + +) do
38      |   if (q == 1) then
39      |   |   set σ_q ← λ_q + ξ.
40      |   else
41      |   |   set σ_q ← λ_q + σ_{q-1}.
42      |   end
43      end
44      Retrieved 8 decimal values of 4 bits binary secret data σ_q where q=1,2,...,8 and received 32 secret
        message bits.
```

---

secret data each, are retrieved from σ_q where q = 1,2, .., 8. Each decimal number provides 4 bits secret message bits. Thus, 32 secret bits (1100 1111 1010 0011 0111 0001 1001 1110) and original bitmap B_i are recovered at the end of the extraction phase. The above extraction technique is applied on all the dual-stego bitmaps AB_i and EB_i sequentially. Finally, all the original cover bitmaps B_i are concatenated including all unmodified pixels to recover the original cover image I. The step-by-step extraction technique is presented in Algorithm 2.

4   Experimental Results, Comparison and Analysis

This section presents the experimental results of the proposed scheme where all the experiments are performed on different standard (512×512) gray-scale and color cover images such as "Lena", "Baboon", "Barbara", "F16", "Lake" and

"Tiffany". The visual quality between cover image (I) and dual-stego images (SI') and (SI'') are determined by Peak Signal-to-Noise Ratio ($PSNR$) (dB), calculated by Eq. 1 in the proposed scheme. High $PSNR$ (dB) value indicates low distortion of stego image.

$$PSNR(x, y) = 10 \log_{10} \left(\frac{max^2}{MSE} \right) \tag{1}$$

where,

max = The highest value of the 8-bit image and MSE = Mean Squared Error.

Another important parameter to estimate the functional efficacy of the proposed scheme is to calculate the embedding capacity which is the maximum number of bits that can be embedded into I. Hence, the maximum bits per pixel (bpp) here is:-

$$\frac{4 \times 8 \times number\ of\ bitmaps}{(512 \times 512)pixel} = \frac{32 \times 512/3 \times 512/3}{(512 \times 512)}\ bpp = 3.56\ bpp$$

The relevant experimental results regarding $PSNR$ (dB) and Embedding Capacity (EC) (in number of bits) of the test images (gray-scale and color) are tabulated in Table 1. To verify the efficiency of the proposed scheme, a comparative study based on the $PSNR$ (dB) and EC of the tested images is made with some relevant state-of-the-art schemes, and is shown in Table 2. In Table 2, it is observed that, the proposed scheme provides 30.49 (db), 12.68 db, 8.94 db and 22.05 db better $PSNR$ values for gray-scale test image $Lena$ and 36.17 db, 18.36 db, 14.62 db and 27.73 db improved $PSNR$ values for color image $Lena$ compared to Lin et al. [9], Pal et al. [10], Jana et al. [11] and Jana et al. [12] schemes respectively. This guarantees better visual quality of the stego images compared to other state-of-the-art schemes. From the same table, it is also noticed that, the proposed scheme allows $46, 110$ bits, $3, 28, 416$ bits and $1, 33, 120$ bits more embedding capacity for the test image $Lena$, which is superior compared to Lin et al. [9], Jana et al. [11] and Jana et al. [12] schemes respectively. But, in this point, the suggested scheme is at per with the scheme of Pal et al. [10], as both of them ensure the same embedding capacity.

Two important parameters of any data hiding scheme are visual quality and embedding capacity. These are inversely proportional to each other. So, for any steganographic scheme, it's a challenge to make a balance between them because higher embedding capacity provides poor visual quality of the stego images. The proposed scheme provides more than 60 dB PSNR value with almost 3,93,000 (bits) embedding capacity for all gray-scale stego images. This trade-off between visual quality (PSNR (dB)) and embedding capacity of the stego images are shown in Table 3. The various sizes of secret images are embedded into the (512×512) different cover images and the changes in $PSNR$ values of the

Table 1. After hiding the secret information, the experimental results of PSNR (dB) and EC (bits) of the gray-scale and color dual-stego images SI' and SI".

Image	Cover Images (512 × 512)	EC (Bits)	PSNR (dB)		
			I vs SI'	I vs SI"	Avg(SI' vs SI")
Gray-scale Images	Lena	3,93,216	63.02	62.77	62.90
	Baboon	3,93,216	62.52	62.43	62.48
	Barbara	3,93,216	64.08	66.80	65.44
	F16	3,93,216	62.80	64.33	63.57
	Lake	3,93,216	62.09	62.87	62.48
	Tiffany	3,93,216	62.11	62.72	62.42
Color Images	Lena	1,179,636	67.73	69.43	68.58
	Baboon	1,179,636	67.32	65.21	66.26
	Barbara	1,179,636	67.08	66.99	67.04
	F16	1,179,636	67.79	69.65	68.72
	Lake	1,179,636	66.69	69.63	68.16
	Tiffany	1,179,636	68.72	67.12	67.92

Table 2. The simulation results comparison between the proposed scheme and some recent state-of-the-art schemes based on two parameters $PSNR$ (dB) and EC $(bits)$ after hiding secret information.

Image	Lin et al. [9]		Pal et al. [10]		Jana et al. [11]		Jana et al. [12]		Proposed (Gray-scale)		Proposed (Color)	
	PSNR (dB)	EC (bits)	PSNR (dB)	EC (bits)	PSNR (dB)	EC (bits)	PSNR (dB)	EC (bits)	PSNR (dB)	EC (bits)	PSNR (dB)	EC (bits)
Lena	32.41	3,47,106	50.22	3,93,216	53.96	64,800	40.85	2,60,096	62.90	3,93,216	68.58	1,179,636
Baboon	26.34	3,52,795	49.97	3,93,216	54.19	64,800	40.82	2,60,096	62.48	3,93,216	66.26	1,179,636
Barbara	31.72	3,30,710	49.91	3,93,216	54.27	64,800	40.87	2,60,096	65.44	3,93,216	67.04	1,179,636
F16	30.33	3,02,000	49.88	3,93,216	53.68	64,800	40.82	2,60,096	63.57	3,93,216	68.72	1,179,636
Lake	29.56	3,38,732	50.01	3,93,216	53.10	64,800	40.75	2,60,096	62.48	3,93,216	68.16	1,179,636
Tiffany	31.60	3,29,028	49.57	3,93,216	53.63	64,800	40.79	2,60,096	62.42	3,93,216	67.92	1,179,636

respective stego images are reflected in Table 3 where from (32 × 32) size to (256 × 256) size secret images are used for embedding. The different sizes of secret images are shown in Fig. 5.

Table 3. Changes in $PSNR$ (dB) values of the various dual stego images SI' and SI" after embedding the different sizes of secret images.

Cover Image (512 × 512)	Changes in PSNR (dB) values after embedding various size of secret images							
	(32 × 32)		(64 × 64)		(128 × 128)		(256 × 256)	
	SI'	SI"	SI'	SI"	SI'	SI"	SI'	SI"
Lena	72.76	71.52	69.84	68.76	66.42	65.65	63.02	62.77
Baboon	71.04	71.84	68.75	68.53	65.85	65.61	62.52	62.43
Barbara	73.63	73.29	71.32	71.06	67.60	68.93	64.08	66.80
F16	71.87	73.63	68.73	70.57	65.59	67.65	62.80	64.33
Lake	70.69	71.49	68.86	69.72	65.48	65.88	62.09	62.87
Tiffany	71.88	72.07	68.69	69.83	65.86	66.09	62.11	62.72

Fig. 5. Various sizes of secret logo applied in the proposed scheme.

4.1   Reversibility Analysis and Histogram Analysis

Figure 6 proves the reversibility of the proposed scheme because for all images the color differences of $(I - RCI)$ image is black with zero pixel intensity which guarantees that original cover image I and Recovered Cover Image (RCI) are exactly identical to each other. Figure 7 displays the histograms of the original cover image $Lena$ (I), respective dual-stego images SI' and SI'' and their respective histogram differences. It is noticed that the histogram differences between (I, SI') and (I, SI'') both are insignificant which assures the good quality of the dual-stego images.

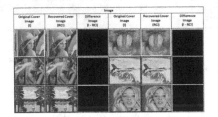

Fig. 6. Reversibility analysis of the proposed scheme by original cover image (I), respective Recovered Cover Image (RCI) and their differences ($I - RCI$).

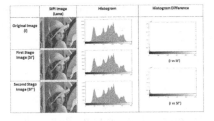

Fig. 7. Histogram of original images I, respective dual-stego images SI' and SI'' and their respective histogram differences.

4.2   Robustness Analysis

To examine the robustness of the proposed algorithm, various types of attacks are performed at different tampering rates such as 10%, 20%, 30%, 40% and 50% on the dual-stego images $Lena$. The $PSNR$ (dB) value of the recovered cover images (RCI) are tabulated in Table 4 at different tampering rates and in most of the cases, it is found that, the $PSNR$ (dB) of the recovered $Lena$ image

Table 4. $PSNR$ (dB) of recovered cover image (RCI) at different tampering rate for the test image *Lena* of proposed the scheme.

Type of Tampered	Tampered on	Tampering Rate				
		10%	20%	30%	40%	50%
Brightening	SI'	59.53	56.34	53.14	49.92	46.86
	SI''	59.04	55.90	52.73	49.21	46.56
	Both	57.12	53.73	50.08	47.23	43.85
Blurring	SI'	58.84	54.98	51.92	47.10	43.89
	SI''	58.24	54.53	50.87	47.74	43.25
	Both	57.11	53.29	49.52	44.43	40.88
Cropping	SI'	36.36	31.27	27.14	22.74	17.89
	SI''	36.24	32.32	28.49	22.06	18.30
	Both	32.72	27.75	23.22	19.62	14.53
Median Filter	SI'	53.45	49.87	45.64	40.73	36.48
	SI''	53.32	48.04	44.33	39.53	36.22
	Both	51.01	46.87	41.39	36.07	32.45
Salt & Pepper	SI'	56.72	53.45	49.94	45.31	41.76
	SI''	57.34	53.38	48.05	44.44	40.59
	Both	53.45	50.61	46.94	42.42	38.42
Contrast	SI'	39.61	34.85	29.69	24.53	20.97
	SI''	38.67	33.24	29.32	24.15	20.70
	Both	36.11	31.53	27.59	22.20	17.75

is more than 20 dB even at 50% tampering level. Hence, it can said that, the proposed approach recovers the cover images at satisfactory level with a high degree of tampering rate, which is gladly acceptable for solving various real life problems, like authentication, copyright protection, ownership identification etc.

5   Conclusion

A unique, efficient and reversible data embedding scheme exploiting Weighted Matrix and Cellular Automata (CA) is proposed here to overcome the loopholes of existing data embedding schemes for image media. Use of weighted matrix and a special class of CA with unity attractors makes this scheme reversible and boosts the security of the proposed scheme as compared to other state-of-the-art schemes. A trade-off between visual quality (PSNR (dB)) and embedding capacity of the stego images put on an extra essence to the proposed scheme. Ownership identification, authentication etc. can be achieved by this proposed scheme, while performing various malicious attacks on the stego images. Hence, this scheme is robust, efficient and useful to fulfil the current market demand. This approach can also be applied in frequency domain in future which may be fruitful to improve the innocuousness and performance of the proposed scheme. Further, in-depth characterization of such special class of CA rules as employed in the current design may also be considered as future extension to strengthen the theoretical background of the proposed scheme.

References

1. Yang, Q., Peng, F., Li, J.-T., Long, M.: Image tamper detection based on noise estimation and lacunarity texture. Multimedia Tools Appl. **75**(17), 10201–10211 (2016)
2. Tralic, D., Grgic, S., Sun, X., Rosin, P.L.: Combining cellular automata and local binary patterns for copy-move forgery detection. Multimedia Tools Appl. **75**(24), 16881–16903 (2016)
3. Al-Shaarani, F., Gutub, A.: Securing matrix counting-based secret-sharing involving crypto steganography. J. King Saud Univ.-Comput. Inf. Sci. **34**(9), 6909–6924 (2022)
4. Priyadharshini, A., Umamaheswari, R., Jayapandian, N., Priyananci, S.: Securing medical images using encryption and LSB steganography. In: 2021 International Conference on Advances in Electrical, Computing, Communication and Sustainable Technologies (ICAECT) CONFERENCE 2021, Bhilai, India, pp. 1–5. IEEE (2021). https://doi.org/10.1109/ICAECT49130.2021.9392396
5. Tafti, A.P., Malakooti, M.V., Ashourian, M., Janosepah, S.: Digital image forgery detection through data embedding in spatial domain and cellular automata. In: The 7th International Conference on Digital Content, Multimedia Technology and its Applications CONFERENCE 2011, Busan, Korea (South), pp. 11–15. IEEE (2011). Electronic ISBN978-89-88678-47-3
6. Kumaresan, G., Gopalan, N.: An analytical study of cellular automata and its applications in cryptography. Int. J. Comput. Network Inf. Secur. **10**(12), 45 (2017)
7. Jayalakshmi, R., Amutha, R.: A theoretical study on the implementation of quantum dot cellular automata. In: 2018 Fourth International Conference on Advances in Electrical, Electronics, Information, Communication and Bio-Informatics (AEE-ICB) CONFERENCE 2018, pp. 1–6. IEEE, Chennai, India (2018). https://doi.org/10.1109/AEEICB.2018.8480984
8. Abdo, A., Lian, S., Ismail, I.A., Amin, M., Diab, H.: A cryptosystem based on elementary cellular automata. Commun. Nonlinear Sci. Numer. Simul. **18**(1), 136–147 (2013)
9. Lin, C.-C., Lin, J., Chang, C.-C.: Reversible data hiding for AMBTC compressed images based on matrix and Hamming coding. Electronics **10**(3), 281 (2021)
10. Pal, P., Jana, B., Bhaumik, J.: Robust watermarking scheme for tamper detection and authentication exploiting CA. IET Image Proc. **13**(12), 2116–2129 (2019)
11. Jana, B., Giri, D., Mondal, S.K.: Dual image based reversible data hiding scheme using (7, 4) hamming code. Multimedia Tools Appl. **77**, 763–785 (2018)
12. Biswapati, J., Debasis, G., Kumar, M.S.: Weighted matrix based reversible data hiding scheme using image interpolation. In: Behera, H.S., Mohapatra, D.P. (eds.) CIDM 2018. AISC, vol. 411, pp. 239–248. Springer, New Delhi (2016). https://doi.org/10.1007/978-81-322-2731-1_22

Characterization of 3-State Cellular Automata with Single Length Cycles

Sukanya Mukherjee[1]([envelope]) and Jakkula Sindhuja[2]

[1] Department of Computer Science and Engineering, Institute of Engineering and Management Kolkata, University of Engineering and Management, Kolkata 700091, West Bengal, India
sukanyaiem@gmail.com

[2] Schlumberger Technology Services India (STS-I), Building-6, Commerzone IT Park Internal Road, Commerzone IT Park, Yerawada, Pune 411006, Maharashtra, India

Abstract. We consider the 3-state 3-neighborhood uniform cellular automata, and study scenarios where the automata can generate one or more *single length cycles*. We show that *self-replicating RMTs* play the key role in this aspect; the greater the number of self-replicating RMTs in a rule, the more is the number of single length cycles generated by a CA with that rule. However, this is not the only criterion; this work identifies a set of *distinguished* RMTs which determines whether the given cellular automaton can generate single length cycle or not. The above analysis can also be used to compute the cycle structures in some special cases. Finally, we demonstrate the efficacy of our approach in pattern classification using real world dataset.

Keywords: single length cycle CA · 3-state CA · self-replicating RMT · classification

1 Introduction

In the configuration space of a finite cellular automaton (CA), the configurations are always attracted to some *cycles* which are also known as *attractors*. A configuration belonging to a cycle is called as *cyclic* configuration; otherwise the configuration is *acyclic*. The *length* of a cycle is determined by the number of configurations involved in that cycle. A cycle with length one is called as *single length cycle* or *point state attractor* or *fixed point*. A CA can have one or more than one single length cycle(s). An automaton with n ($\in \mathbb{N}$) cells can be called as *single length cycle* cellular automaton if it generates point state attractor(s) only. This type of cellular automata (CAs) has gained popularity due to its *convergence* property i.e., here the configurations move towards one or more fixed point attractors and makes the system *stable*. The journey of study

This work is carried out as a project in the Summer School on Cellular Automata Technology 2023.

of convergence property started with Wolfram's class I rules in *uniform* CAs (in a uniform CA, all cells follow same rule) [21]. The inclusion of non-uniformity in rule selection for CA cells added more challenges, and convergence property of special class of Wolfram's cellular automata, namely the linear/additive CAs, has been studied in [4, 14]. Detailed discussion on single length cycle cellular automata in the domain of non-linear CAs is presented in [1, 7, 19]. A single length cycle cellular automaton can be categorized as either single attractor cellular automata (SACA) or multi attractor cellular automata (MACA) based on the count of the point state attractors it owns; the former refers to the CA where the count of cycle is one and the later one has no restriction on the number of cycles. SACA [3, 4, 6, 12, 13] and MACA [1, 4, 7, 14, 17, 19] have been explored extensively. Besides theoretical aspects, CAs with single length cycles are useful to detect faulty nodes in WSN [8], associate memory design [2, 5, 9, 11], and are highly demanding in the domain of pattern classification [10, 15–18, 20].

Prior art on single length cycle cellular automata have explored only Wolfram's CAs or Elementary CAs (ECAs); for these type of automata, CA cells maintain 3-neighborhood architecture and cell state value is either 0 and 1. It is challenging to extend the study of the convergence property to CAs where the count of states is more than two or neighborhood structure is more than three. Our work focuses on the characterization of uniform 3-state single length cycle cellular automata (where the cells states are 0, 1 and 2), keeping neighborhood structure same as of prior art; this study can be a good starting point to explore the more general case of d-state ($d \geq 2$) single length cycle cellular automata, for arbitrary values of d. As ECAs has been successfully applied in pattern classification, we are motivated to study the performance of 3-state single length cycle cellular automata in pattern classification.

The generation of next configuration from the current one is always governed by a *Rule Min Term (RMT) sequence*. An RMT represents the combination of a cell's present state y along with those (x and z) of its neighboring cells; for the current scenario, an RMT can be viewed as xyz where $x, y, z \in \{0, 1, 2\}$. An RMT can also be represented as r which is the decimal equivalent of xyz. An RMT sequence is actually an n-length string of the decimal equivalents of RMTs when the given CA uses n cells. As this work uses 3-state 3-neighborhood architecture, therefore, we can get $\{0, 1, 2\}^3 = 27$ possible RMTs.

An RMT xyz or r is termed *self-replicating* if the next state w.r.t the given RMT is y, i.e., there is no change in the cell state. Hence, to get a point state attractor, all the RMTs of the corresponding RMT sequence must be self-replicating. Intuitively, it can be believed that greater the number of self-replicating RMTs in a rule, more is the chance of the corresponding CA to yield single length cycle(s) only; however, this is not always true; e.g., a 3-state CA rule with two non self-replicating RMTs (002 and 012 with next state values 1 and 0 respectively) and all remaining 25 self-replicating RMTs can never generate a single length cycle cellular automaton. This motivates us to identify a collection of self-replicating RMTs that guarantees a CA to be a single length cycle cellular automaton. Additionally, a collection of rules is identified, the presence of a member of which restrains the CA to generate any single length cycle.

This work demonstrates an experimental study on generation of single length cycle cellular automata by reducing the number of self-replicating RMTs in an unique style. By the experimental study, we can get not only the single length cycle cellular automata but also their *cycle structure* for a collection of CA sizes, for a specific rule set. All the above mentioned study related to the characterization of 3-state 3-neighborhood single length cycle cellular automata is reported in Sect. 3. The efficacy of 3-state single length cycle CA based pattern classifier is reported in Sect. 4. For designing classifier, those single length cycle CAs are considered which can generate *moderate* number of cycles. Here, 3-state CA based classifier starts performing on real dataset which consider *quantitative* attributes only. As the first step of classification, the real data are to be mapped to CA configurations using an encoding technique. We use real data from UCI Machine Learning Repository. Next section explains some terminologies and notations of 3-state cellular automata useful for our work.

2   3-State Cellular Automata

A 1-dimensional cellular automata of *size* n consists of n *cells*, numbered 0 to $n-1$. Equivalently, we call such CA as n-cell CA. Each cell contains a *state*, and let S denote the set of possible states of the cells. In this work, we focus on 3-state cellular automata; thus $S = \{0, 1, 2\}$. The sequence of the state values of all n cells together is called as *configuration*. Let x be a configuration of n-cell CA. Thus x can be represented as $(x_0 x_1 \cdots x_{n-1})$ where every x_i belongs to S; a configuration can also be represented as the decimal equivalent of $(x_0 x_1 \cdots x_{n-1})$. Let C be the configuration space of an n-cell 3-state CA; therefore, $C = \{0, 1, 2\}^n$.

Traditionally, at every time step, all the CA cells update their state values together and from the current configuration, we get the *next configuration*. Now, the update of a cell is influenced by its neighbors. As this work supports 3-neighborhood architecture, to get the *next state* of cell i, we consider the *present states* of cell $i-1$ (left neighbor of cell i), the state of the cell itself and the state of its right neighbor i.e., cell $i+1$. Let $\mathcal{R}$ be the *next state function* such that $\mathcal{R} : \{0, 1, 2\}^3 \rightarrow \{0, 1, 2\}$. Thus $\mathcal{R}(x_{i-1}, x_i, x_{i+1}) = y_i$ where y_i is next state of cell i and x_{i-1}, x_i and x_{i+1} are the present states of cells $i-1$, i and $i+1$ respectively. If a CA follows $m = (r_l + r_r + 1)$-neighborhood architecture where r_l is the count of left neighbors and r_r is the count of right neighbors, then the sequences of the present states from cells $i - r_l$ to $i-1$, cell i and cells $i+1$ to $i+r_r$ is called as an *Rule Min Term* or RMT for cell i. In this work, a string $x_{i-1} x_i x_{i+1}$ of length three refers to an RMT as $m = 1 + 1 + 1 = 3$ where $r_l = r_r = 1$. Thus in our work, $3^3 = 27$ RMTs can be possible. An RMT can also be viewed as the decimal equivalent of $x_{i-1} x_i x_{i+1}$. Let r be the decimal representation of $x_{i-1} x_i x_{i+1}$; therefore, $r = 3^2.x_{i-1} + 3^1.x_i + 3^0.x_{i+1}$. In this way, here, $\mathcal{R}(x_{i-1}, x_i, x_{i+1})$ or equivalently $\mathcal{R}[r]$ refers to either 0 or 1 or 2. Thus $\mathcal{R}$

can be represented as a string of length 27 and constituted by $\{0, 1, 2\}$; this $\mathcal{R}$ can also be represented as the decimal equivalent of that string. Here, $\mathcal{R}$ is called as the *rule* applied to the cell i. Thus we get, $\{0, 1, 2\}^{27}$ possible rules for 3-states 3-neighborhood cellular automata. Here, Table 1 describes the representation of 3-state CA rules. As we deal with CA with finite number of cells, so, we need to consider some boundary condition. In this work, *null* boundary cellular automaton is considered. Therefore, to update cell 0, only *nine* RMTs from 27 RMTs are considered as $x_{-1} = 0$. These nine RMTs are $000(0), 001(1), 002(2),$ $010(3), 011(4), 012(5), 020(6), 021(7)$ and $022(8)$. Similarly, for cell $n - 1$, again, *nine* RMTs are useful as $x_n = 0$ and these are as follows: $000(0), 010(3), 020(6),$ $100(9), 110(12), 120(15), 200(18), 210(21)$ and $220(24)$.

Table 1. A very few 3-state CA rules. Here, PS refers to the sequence of the present states of the left neighbor of the given cell, the cell itself and the cell's right neighbor. This PS is equivalent to RMT and (RMT) refers to the decimal equivalent of PS. NS is the next state value for the given RMT.

PS	222	221	220	212	211	210	202	201	200	122	121	120	112	111	110	102	101	100	022	021	020	012	011	010	002	001	000
(RMT)	(26)	(25)	(24)	(23)	(22)	(21)	(20)	(19)	(18)	(17)	(16)	(15)	(14)	(13)	(12)	(11)	(10)	(9)	(8)	(7)	(6)	(5)	(4)	(3)	(2)	(1)	(0)
NS	1	0	1	2	0	0	1	1	1	1	0	0	2	2	0	2	1	2	1	0	1	2	2	2	2	2	2
	2	2	2	1	1	1	0	0	0	2	2	2	1	1	1	0	0	0	2	2	2	1	1	1	0	0	0
	2	2	2	1	1	1	0	0	0	2	2	2	1	1	1	0	0	0	2	2	2	0	1	1	1	0	0
	1	0	1	2	0	0	1	1	1	1	0	0	2	2	0	2	1	2	1	0	1	2	2	2	2	2	0

Table 2. Here, r_i and r_{i+1} are the RMTs at two consecutive cells i and $i + 1$.

r_i	0/9/18	1/10/19	2/11/20	3/12/21	4/13/22	5/14/23	6/15/24	7/16/25	8/17/26
r_{i+1}	0, 1, 2	3, 4, 5	6, 7, 8	9, 10, 11	12, 13, 14	15, 16, 17	18, 19, 20	21, 22, 23	24, 25, 26

As this work considers uniform cellular automata, therefore, all the cells follow the same rule $\mathcal{R}$. Therefore, we can get $3^{27} = 7.625597485 \times 10^{12}$ cellular automata. An n-cell CA can be thought of as function G_n such that $G_n : C \rightarrow C$. Let $\mathbf{x} = (x_0 x_1 \cdots x_{n-1})$ and $\mathbf{y} = (y_0 y_1 \cdots y_{n-1})$ be two configurations of an n-cell 3-state CA. If $G_n(\mathbf{x}) = \mathbf{y}$ where every y_i can be computed as follows: $y_i = \mathcal{R}(x_{i-1}, x_i, x_{i+1})$ provided $0 \le i \le n - 1$, then $\mathbf{y}$ is *reachable* from $\mathbf{x}$ and $\mathbf{x}$ is the *predecessor* of $\mathbf{y}$. If $G_n^l(\mathbf{x}) = \mathbf{x}$ for a finite l, then $\mathbf{x}$ is called as *cyclic* configuration and forms a *cycle*; otherwise, $\mathbf{x}$ is *acyclic*. When such l is 1, then x forms *single length cycle* or *point state attractor* or *fixed point*. When l is greater than one, then the corresponding cycle is called as *multi-length cycle* or *multi-state attractor*. Here, l is the *length* of a cycle i.e., the number of configurations involved in that cycle. The configuration space of finite CA can also represent its *cyclic space*. The cyclic space of a CA refers to the cyclic and acyclic configurations, number of cycles it has and their lengths also. The *cycle structure* (CS) of a CA represents the collection of *cyclic components* $\mu_i(l_i)$s where each μ_i is the number of cycles having the distinct length l_i. Therefore, the cycle structure of a CA $\mathcal{R}$ can be represented as $CS_\mathcal{R} = [\mu_1(l_1), \mu_2(l_2), \cdots]$. For single length cycle CA, $CS_\mathcal{R} = [\mu(1)]$. In the cyclic space of a CA, if all configurations are cyclic, then such CA is called as *reversible* CA; otherwise CA is *irreversible*. In an irreversible CA, there exist cyclic and acyclic configurations both. An acyclic configuration which is not reachable from any other configuration of the CA, then such acyclic configuration is called as *non-reachable* configuration. This type of configuration has no predecessor. Let us explain the above mentioned terminologies by the help of a 3-state cellular automata $\mathcal{R} = 111222000000000000000000000$ (decimal equivalent: $3.94355317E12$) having 3 cells. In this CA, out of 3^3 configurations, one configuration (0) is cyclic and all the remaining configurations are acyclic. Therefore, the corresponding CA is irreversible. Out of 26 acyclic configurations, 22 configurations are non-reachable. Here, $CS_\mathcal{R} = [1(1)]$; therefore, $\mathcal{R}$ is single length cycle CA.

It has already been discussed that the change of the state of a cell is determined by the *value* at the corresponding RMT. For example, if we look at row 5 of Table 1, we get the value at RMT 222 is 2 for the given rule of row 5. Let a cell maintain RMT 222; therefore, its present state and next state are same i.e., 2; whereas, by applying the rule of row 6, if a cell maintains RMT 222, that cell has different next state with respect to the present state because the value at 222 is 1. An RMT xyz ($x, y, z \in \{0, 1, 2\}$) of a rule $\mathcal{R}$ is said to be *self-replicating* if the value at RMT xyz is y. In the prior examples, the former one refers to RMT 222 as self-replicating RMT whereas the later one does not. A reachable configuration can also be viewed as a sequence of RMTs. Let us consider a 3-cell CA $\mathcal{R} = 1012001111002202121010222222$ (row 3 of Table 1). If the current configuration of the CA is 012, then the next configuration can be found as follows as we consider null boundary CA: $\mathcal{R}(0, 0, 1)\mathcal{R}(0, 1, 2)\mathcal{R}(1, 2, 0) = \mathcal{R}[1]\mathcal{R}[5]\mathcal{R}[15] = 220$. Here, $(1, 5, 15)$ is called as the *RMT sequence* corresponding to the configuration 220. A configuration is to be in a single length cycle if the corresponding RMT sequence maintains self-replicating RMTs for every cell (each RMT of RMT sequence $(0, 0, 0)$ of CA $111222000000000000000000000$ is self-replicating and generates a single length cycle using the configuration 000). Next, we present on the generation of RMT sequence. Let r be the decimal equivalent of RMT xyz. As we consider 3-state CA, therefore, if the current cell i maintains RMT xyz, then the next cell will have the RMT as either $yz0$ or $yz1$ or $yz2$. Now, x can be either 0 or 1 or 2. For example, if the RMT for cell i is any one from $\{000, 100, 200\}$, then the RMTs for cell $i + 1$ will be from $\{000, 001, 002\}$; only the constraint is imposed for the last cell as we consider null boundary CA and for this example, such RMT is 000 only. Table 2 represents which RMTs can be placed at the consecutive positions in a RMT sequence.

Next, we introduce two terminologies on RMTs. The RMTs $xy0$, $xy1$ and $xy2$ are said to be *sibling* and the RMTs $0yz$, $1yz$ and $2yz$ are said to be *equivalent* where $x, y, z \in \{0, 1, 2\}$. For 3-state CA, we get *nine* different sets of sibling RMTs and *nine* more different sets for equivalent RMTs. For example, $\{0, 1, 2\}$ is a set of sibling RMTs and $\{0, 9, 18\}$ is a set of equivalent RMTs.

3   Characterization of 3-State CA with Single Length Cycle only

RMTs play the key role in determining of 3-state 3-neighborhood single length cycle cellular automata. Hereafter, we call 3-state CA with 3-neighborhood architecture as 3-state CA. Out of 3^{27} 3-state CAs, there is only one CA 222111000222111000222111000, which is single length cycle CA for any given CA size; the reason behind is that all the 27 RMTs are self-replicating. For this CA, the cycle structure is $[2^n(1)]$ where n is the CA size. Now a CA generating single length cycle must have some self-replicating RMTs in the corresponding rule but the existence of self-replicating RMTs in CA rule does not guarantee the generation of single length cycle. For example, 3-state CA with rule 0^{27} of an arbitrary CA size n, always generates a single length cycle by the configuration 0^n as the corresponding RMT sequence $(0, 0, \cdots, 0)$ confirms self-replicating RMT 0 for every cell position. On contrary, CA 222111000222111000101220122 of size n with null boundary condition can never form single length cycle for any n, though the corresponding CA rule maintains 18(66.7%) self-replicating RMTs. As we consider null boundary CA, therefore, for cell 0, only RMTs 0 to 8 will be considered and none of them are self-replicating for the given rule. Therefore, we can never get any single length cycle as there is no existence of any RMT sequence which maintains all self-replicating RMTs. This example even establishes that higher number of self-replicating RMTs is not the only requirement for getting single length cycle cellular automata; rather we need to investigate the intrinsic properties of RMTs which guide us to figure out the CAs with only single length cycles.

3.1   Insignificant Rules

When we explore 3-state CAs, we may get some CAs which form single length cycles only, some CAs with single length and multi-length cycles and the remaining CAs form multi-length cycles only. To identify CAs with single length cycles only from 3^{27} CAs, classically, we need to find the cycle structures of these CAs which are infeasible in practice. In a CA, the number of single length cycles can be found in linear time but that does not guarantee whether such CA forms a multi-length cycle or not. So, our first target is to find those 3-state CAs which can never form any single length cycle for an arbitrary CA size. For this investigation, self-replicating property of RMTs drives us.

Lemma 1. *An n-cell CA can never generate a single length cycle if there does not exist any RMT sequence $(r_0, r_1, \cdots, r_{n-1})$ where every r_i is self-replicating.*

Proof. For an n-cell CA, let $(r_0, r_1, \cdots, r_{n-1})$ represent an RMT sequence. Let us consider the given scenario. Let $r_0, r_1, \cdots, r_{i-1}$ be the self-replicating RMTs and r_i be a non self-replicating RMT. Suppose $r_i = xyz$ where $x, y, z \in S$. Now, if we get a configuration which contains xyz, as r_i is non self-replicating, therefore, in the next configuration at the place of state y, we get such state which is from $S \setminus \{y\}$. Therefore, the current configuration and its next configuration do not contain same state at least for one cell. Thus, by $(r_0, r_1, \cdots, r_{n-1})$, we do not get any single length cycle. Now, for each of the remaining $d^n - 1$ (where d is the number of states and in this work, $d = 3$) RMT sequences, if there exists at least one RMT which is non self-replicating, then the CA can never generate a single length cycle. Hence, the proof follows.   □

Let us devise our *first* strategy for discarding insignificant rules. Lemma 1 implies that if all possible RMTs in a rule are not self-replicating, then that rule can be discarded from our target rule space to study CAs with single length cycles only. A 3-state CA constructed by $\mathcal{R}$ can never generate a single length cycle if the following condition holds: each of $\mathcal{R}[x0z]$ is either 1 or 2, each of $\mathcal{R}[x1z]$ is either 0 or 2 and each of $\mathcal{R}[x2z]$ is either 0 or 1 where $x, z \in \{0, 1, 2\}$. Thus we can get total $2^9 \cdot 2^9 \cdot 2^9 = 2^{27}$ 3-state CAs that form only *multi-length cycle(s)*. Let R_1 denote the set of those 2^{27} rules. Suppose R denote the rule space for 3-state CAs; then we need to discard the rules of R_1 from R. Additionally, as discussed before, even if a rule maintains a large number of self-replicating RMTs (for example 66.7%), it may not generate a single length cycle. So, let us discuss our *second* strategy for finding the insignificant rules for CAs with single length cycles only. In this strategy, some of the RMTs can be self-replicating.

Here, cell 0 of an n-cell CA always maintains RMT $0yz$ where $y, z \in \{0, 1, 2\}$. To get at least one single length cycle in a CA, one RMT sequence should be designed in such a way that each of the RMT in that sequence is self-replicating. If all RMTs in the form of $0yz$ are non self-replicating without any restrictions on the RMTs of the forms $1yz$ and $2yz$, then the corresponding CA can never form even one single length cycle. If all the RMTs of the form of $0yz$ are non self-replicating, then we can get 2^9 combinations where the next states of RMTs of the form $0yz$ are from $\{0, 1, 2\} \setminus \{y\}$. As there is no restriction on the next state selection for the RMTs of the forms $1yz$ and $2yz$, therefore, we get total $(3^9 \cdot 3^9) \cdot 2^9 = 3^{18} \cdot 2^9$ rules and let R_2 denote the set of those rules. Thus these $3^{18} \cdot 2^9$ rules also can never form single length cycle CAs. Here, every rule of R_2 maintains at least 33.3% non self-replicating RMTs; even if we maintain 33.3% arbitrary non self-replicating RMTs in a rule, it can not guarantee that the CA rule generates multi-length cycles only. For example, in a rule if RMTs 9(100) to 17(122) are non self-replicating and all remaining are self-replicating RMTs, then an n-cell CA constructed by that rule can generate single length cycles (here, 0^n and 2^n are two configurations which form single length cycles as RMTs 000(0), 022(8) and 222(26) are self-replicating). Next, we show that even if we increase the number of self-replicating RMTs strategically, the CAs may produce multi-length cycles only.

Let us focus on first 9 RMTs (000(0) to 022(8)). We restrict the first 3 RMTs 000(0), 001(1) and 002(2) to be non self-replicating. Let us consider the case when the count of self-replicating RMT is *one* from RMTs 3 to 8. Suppose RMT $0yz$ be the only RMT which is self-replicating. Now, if RMTs $yz0$, $yz1$ and $yz2$ are non self-replicating, then there is no way of getting any single length cycle in a CA. Let us calculate the number of such rules. Out 9 RMTs, when one is self-replicating and remaining 8 are non self-replicating, we get 2^8 possible distinct combinations. Moreover, as 3 RMTs from RMT 100(9) to 222(26) need to be non self-replicating, so, we can get 2^8 distinct such combinations. Now, at the remaining 15 RMTs, the next states can be any of $\{0, 1, 2\}$. Thus we get, total $\binom{6}{1} \cdot 2^8 \cdot 2^3 \cdot 3^{15}$ such rules which follow one self-replicating RMT from RMTs 3 to 8. In general, we can get k self-replicating RMTs between RMTs 3 to 8, where k lies between 1 to 6. Recall that RMTs 0 to 2 are non self-replicating. Thus the number of CAs which can produce only multi-length cycles where the possible next RMTs (within the range of 9 to 26) of the self-replicating RMTs (from 3 to 8) are non self-replicating is $\sum_{k=1}^{6} \binom{6}{k} \cdot 1^k \cdot 2^{9-k} \cdot 2^{3k} \cdot 3^{18-3k}$. Let R_3 denote the corresponding rule space. Therefore, the CAs of arbitrary sizes always generate multi length cycles only if the corresponding rules are from $(R_1 \cup R_2 \cup R_3)$.

Definition 1. *Let us consider a rule $\mathcal{R}$ for designing an n-cell CA. If the CA always generate multi-length cycles only for an arbitrary n, then $\mathcal{R}$ is called as strictly multi-length cycle rule. A CA constructed by strictly multi-length cycle rule, is called as strictly multi-length cycle CA.*

Therefore, any rule $\mathcal{R} \in (R_1 \cup R_2 \cup R_3)$ is *strictly multi-length cycle* rule and *insignificant* for generating an 3-state CA with single length cycle only.

3.2 Effective Non Self-replicating RMTs for Single Length Cycle CAs

In this section we mainly focus on identification of CAs with single length cycles only. For that, first, we exclude $(R_1 \cup R_2 \cup R_3)$ from R. From the prior discussion, a CA rule with more self-replicating RMTs (for example, 18 out of 27) can generate strictly multi length cycle CA. This instigates us to study the effectiveness of the arrangement of self-replicating and non self-replicating RMTs in a rule for generating single length cycle CA.

Definition 2. *For a given rule $\mathcal{R}$, if an n-cell CA always generates single length cycles only, for an arbitrary n, then $\mathcal{R}$ is called as strictly single length cycle rule and the corresponding CA is called as strictly single length cycle CA.*

Next, we figure out *strictly single length cycle rules* only from $R \setminus (R_1 \cup R_2 \cup R_3)$.

Proposition 1. *In a rule $\mathcal{R}$, if the non self-replicating RMT(s) is(are) from only $\{x0z | x, z \in \{0, 1, 2\}\}$, then $\mathcal{R}$ is strictly single length cycle rule.*

Proof. Let us consider a RMT $r = x0z$ of a rule $\mathcal{R}$ where $x, z \in \{0, 1, 2\}$. Let r be $000(0)$ specifically. Suppose, r is non self-replicating i.e. $\mathcal{R}[0]$ is either 1 or 2. Now, suppose that all the non self-replicating RMTs are of the form $x0z$. If the configuration at time step t is given as $(\cdots x_{i-1}x_i x_{i+1} \cdots) = (\cdots 000 \cdots)$, therefore, the configuration at $t + 1$ will be any one of the following forms: $(\cdots y_i \cdots) = (\cdots 1 \cdots)$ and $(\cdots y_i \cdots) = (\cdots 2 \cdots)$. Now, the configuration at time step $t + 2$ will never maintain 0 at cell i as the RMTs in the forms of $x1z$ and $x2z$ are self-replicating. So, there is no possibility of coming back of configuration $(\cdots x_{i-1}x_i x_{i+1} \cdots) = (\cdots 000 \cdots)$ after two time steps. Moreover, the configuration $(\cdots x_{i-1}x_i x_{i+1} \cdots) = (\cdots 000 \cdots)$ can never return even after two steps. The reason is as follows: cell i should have state 0 after $t + k$ time steps but cell i will maintain state 1 (resp. 2) in the next configuration(s) until the point state attractor is reached. Thus if $\mathcal{R}$ maintains $x0z$ as non self-replicating RMTs at least for one x and z, then $\mathcal{R}$ is strictly single length cycle rule.

Proposition 2. *In a rule $\mathcal{R}$, if the non self-replicating RMT(s) is(are) from only $\{x1z | x, z \in \{0, 1, 2\}\}$, then $\mathcal{R}$ is strictly single length cycle rule.*

Proof. Analogous to the proof of Proposition 1.

Proposition 3. *In a rule $\mathcal{R}$, if the non self-replicating RMT(s) is(are) from only $\{x2z | x, z \in \{0, 1, 2\}\}$, then $\mathcal{R}$ is strictly single length cycle rule.*

Proof. Analogous to the proof of Proposition 1.

From Proposition 1, we can deduce the count of strictly single length cycle rules. In this case, 18 RMTs of the forms $x1z$ and $x2z$ are always self-replicating. When only one RMT out of 9 RMTs of the form $x0z$ is non self-replicating, then we can get $\binom{9}{1} \cdot 2^1$ strictly single length cycle rules. Similarly, out of 9 RMTs, we can get 2, 3, $\cdots$ and even all 9 RMTs as non self-replicating. Thus we get total $\binom{9}{1} \cdot 2^1 + \binom{9}{2} \cdot 2^2 + \cdots + \binom{9}{9} \cdot 2^9 = (1 + 2)^9 - 1 = 3^9 - 1 = 19682$ strictly single length cycle rules. In an analogous manner (using Proposition 2 and Proposition 3), we get $19682 + 19682 = 39364$ more strictly single length cycle rules. Therefore, we get total $3 \cdot 19682 = 59046$ *strictly single length cycle CAs*. Out of these 59046 strictly single length cycle CAs, we explore 19682 such CAs where the RMTs of the form $x0z$ maintain self-replicating property, for figuring out their cycle structures. The purpose is to look on how the number of self-replicating RMTs affects the number of single length cycles.

Let A_0 denote the set $\{000(0), 001(1), 002(2), 100(9), 101(10), 102(11), 200$ $(18), 201(19), 202(20)\}$. In our experiments, we change the count of non self-replicating RMTs in A_0. When number of non self-replicating RMT is *one*, then we get only $1^9 \cdot 1^9 \cdot \binom{9}{1} \cdot 2^1 = 18$ strictly single length cycle CAs as 9 RMTs of the form $x1z$ and 9 RMTs of the form $x2z$ present 1 and 2 as the next state values respectively. Similarly, when *two* RMTs are non self-replicating, then $1^9 \cdot 1^9 \cdot \binom{9}{2} \cdot 2^2 = 144$ CAs are strictly single length cycle CAs. Thus we get 512 such CAs when number non self-replicating RMTs is *nine*. So, we classify 19682 strictly single length cycle rules in *nine* different sets based on the number of non self-replicating they have from the set A_0. Let R_0^i denote the set of strictly single length cycle rules where the rules maintain i non self-replicating RMTs from set A_0. Here, we have chosen the CA sizes from 4 to 7 for every R_0^i ($1 \leq i \leq 9$) to compute the cycle structure of each rule belongs to a given R_0^i.

Table 3. The grouping of strictly single length cycle rules of R_0^1 based on their cycle structures for the CA sizes 4 to 7

# of strictly single length cycle rules in R_0^1	# of the strictly single length cycle CA	CS when $n = 4$	CS when $n = 5$	CS when $n = 6$	CS when $n = 7$
18	2	[64(1)]	[188(1)]	[548(1)]	[1600(1)]
	8	[66(1)]	[190(1)]	[547(1)]	[1575(1)]
	8	[75(1)]	[217(1)]	[628(1)]	[1817(1)]

Table 4. The grouping of strictly single length cycle rules of R_0^2 based on their cycle structures for the CA sizes 4 to 7

# of strictly single length cycle rules in R_0^2	# of the strictly single length cycle CA	CS when $n = 4$	CS when $n = 5$	CS when $n = 6$	CS when $n = 7$
144	8	[51(1)]	[139(1)]	[379(1)]	[1035(1)]
	8	[53(1)]	[149(1)]	[419(1)]	[1177(1)]
	16	[54(1)]	[153(1)]	[433(1)]	[1226(1)]
	8	[55(1)]	[154(1)]	[432(1)]	[1214(1)]
	8	[58(1)]	[163(1)]	[453(1)]	[1261(1)]
	8	[58(1)]	[164(1)]	[459(1)]	[1285(1)]
	16	[60(1)]	[164(1)]	[448(1)]	[1224(1)]
	16	[60(1)]	[165(1)]	[454(1)]	[1249(1)]
	16	[61(1)]	[168(1)]	[463(1)]	[1275(1)]
	16	[61(1)]	[169(1)]	[468(1)]	[1296(1)]
	24	[69(1)]	[191(1)]	[529(1)]	[1465(1)]

Table 3, 4, 5, 6, 7 and Table 8 represent the cycle structures of strictly single length cycle CAs which are constructed by the non self-replicating RMTs from all different R_0^i where $i = 1, 2, 3, 4, 8, 9$. The column I of each of these tables refers to the cardinality of R_0^i. Columns III to VI of Table 3, 4, 5, 6, 7 and Table 8 refer to the cycle structures of the CAs generated by R_0^i. Each entry of column II refers to the count of those strictly single length cycle CAs which have same cycle structure for a given CA size. This work also presents the distribution of configurations as single length cycles. Table 9 represents the relationship between the number of non self-replicating RMTs and the CA sizes (4 to 7). Columns II to V of Table 9 describes the minimum as well as maximum percentage of

Table 5. The grouping of strictly single length cycle rules of R_0^3 based on their cycle structures for the CA sizes 4 to 7

# of strictly single length cycle rules in R_0^3	# of the strictly single length cycle CA	CS when $n = 4$	CS when $n = 5$	CS when $n = 6$	CS when $n = 7$
672	32	[42(1)]	[114(1)]	[310(1)]	[846(1)]
	16	[44(1)]	[120(1)]	[328(1)]	[896(1)]
	32	[46(1)]	[118(1)]	[303(1)]	[778(1)]
	32	[46(1)]	[119(1)]	[308(1)]	[798(1)]
	16	[46(1)]	[127(1)]	[354(1)]	[984(1)]
	16	[46(1)]	[129(1)]	[359(1)]	[999(1)]
	16	[47(1)]	[124(1)]	[328(1)]	[868(1)]
	32	[48(1)]	[129(1)]	[345(1)]	[923(1)]
	32	[48(1)]	[129(1)]	[346(1)]	[927(1)]
	32	[48(1)]	[130(1)]	[351(1)]	[946(1)]
	16	[49(1)]	[130(1)]	[347(1)]	[927(1)]
	16	[49(1)]	[131(1)]	[352(1)]	[941(1)]
	32	[49(1)]	[132(1)]	[355(1)]	[956(1)]
	32	[40(1)]	[133(1)]	[360(1)]	[976(1)]
	32	[50(1)]	[133(1)]	[354(1)]	[943(1)]
	32	[51(1)]	[137(1)]	[369(1)]	[995(1)]
	48	[52(1)]	[140(1)]	[372(1)]	[988(1)]
	32	[54(1)]	[140(1)]	[363(1)]	[941(1)]
	128	[55(1)]	[144(1)]	[377(1)]	[987(1)]
	32	[56(1)]	[148(1)]	[391(1)]	[1033(1)]
	32	[63(1)]	[166(1)]	[438(1)]	[1156(1)]
	32	[63(1)]	[167(1)]	[444(1)]	[1180(1)]

configurations involved in single length cycles. For example, when the count of non self-replicating is 1 and CA size (n) is 4 (see row 2 of the table), out of 18 strictly single length cycle CAs, some of the CAs maintain 79% configurations as single length cycles and another set of CAs form single length cycles with 92.5% configurations. It is depicted from the table that the count of non self-replicating RMTs is *inversely proportional* to the count of single length cycles by keeping the CA size n fixed. When the count of non self-replicating RMTs is fixed then also number of single length cycles reduce gradually with respect to the increase of CA size.

When we use non self-replicating RMTs from A_0, we get only strictly single length cycle CAs. In this regard, the maximum number of non self-replicating RMTs is restricted to 9. It is interesting to check if we get strictly cycle length CAs when we consider more than such 9 non self-replicating RMTs. Let A_1 and A_2 denote the sets of RMTs of the forms of $x1z$ and $x2z$ respectively; $A_1 = \{010(3), 011(4), 012(5), 110\ (12), 111(13), 112(14), 210(21), 211(22), 212(23)\}$ and

Table 6. The grouping of strictly single length cycle rules of R_0^4 based on their cycle structures for the CA sizes 4 to 7

# of strictly single length cycle rules in R_0^4	# of the strictly single length cycle CA	CS when $n = 4$	CS when $n = 5$	CS when $n = 6$	CS when $n = 7$
2016	16	[33(1)]	[89(1)]	[241(1)]	[657(1)]
	64	[37(1)]	[94(1)]	[240(1)]	[615(1)]
	64	[37(1)]	[95(1)]	[245(1)]	[634(1)]
	64	[38(1)]	[97(1)]	[249(1)]	[639(1)]
	64	[38(1)]	[98(1)]	[253(1)]	[655(1)]
	64	[38(1)]	[104(1)]	[284(1)]	[776(1)]
	64	[39(1)]	[100(1)]	[257(1)]	[660(1)]
	64	[39(1)]	[101(1)]	[262(1)]	[679(1)]
	32	[40(1)]	[104(1)]	[273(1)]	[714(1)]
	32	[40(1)]	[107(1)]	[284(1)]	[749(1)]
	192	[41(1)]	[99(1)]	[239(1)]	[577(1)]
	64	[41(1)]	[108(1)]	[286(1)]	[757(1)]
	64	[41(1)]	[109(1)]	[286(1)]	[752(1)]
	64	[41(1)]	[105(1)]	[263(1)]	[659(1)]
	64	[41(1)]	[107(1)]	[271(1)]	[784(1)]
	32	[41(1)]	[109(1)]	[289(1)]	[760(1)]
	32	[41(1)]	[112(1)]	[298(1)]	[792(1)]
	64	[42(1)]	[105(1)]	[263(1)]	[659(1)]
	64	[42(1)]	[107(1)]	[271(1)]	[684(1)]
	32	[42(1)]	[109(1)]	[289(1)]	[760(1)]
	32	[42(1)]	[112(1)]	[298(1)]	[792(1)]

Table 7. The grouping of strictly single length cycle rules of R_0^8 based on their cycle structures for the CA sizes 4 to 7

# of strictly single length cycle rules in R_0^8	# of the strictly single length cycle CA	CS when $n = 4$	CS when $n = 5$	CS when $n = 6$	CS when $n = 7$
2304	256	[17(1)]	[33(1)]	[65(1)]	[129(1)]
	1024	[20(1)]	[40(1)]	[80(1)]	[160(1)]
	512	[20(1)]	[44(1)]	[97(1)]	[214(1)]
	512	[20(1)]	[45(1)]	[101(1)]	[227(1)]

Table 8. The grouping of strictly single length cycle rules of R_0^9 based on their cycle structures for the CA sizes 4 to 7

# of strictly single length cycle rules in R_0^9	# of the strictly single length cycle CA	CS when $n = 4$	CS when $n = 5$	CS when $n = 6$	CS when $n = 7$
512	512	[16(1)]	[32(1)]	[64(1)]	[128(1)]

Table 9. Significance of the proportion of non self-replicating RMTs from A_0 and the CA size in the count of single length cycles

# of selected non self replicating RMTs from A_0	$n = 4$	$n = 5$	$n = 6$	$n = 7$
1	79% and 92.5%	77.3% and 89.3%	75% and 86.1%	72% and 83%
2	62.9% and 85.2%	57.2% and 78.6%	51.9% and 72.5%	47.3% and 66.9%
3	51.8% and 77.7%	46.9% and 68.7%	41.5% and 60.9%	35.3% and 53.9%
4	40.7% and 70.3%	36.6% and 58.8%	32.7% and 49.5%	26.3% and 41.6%
5	35.8% and 54.3%	30% and 41.9%	24.2% and 32.9%	17.8% and 29.9%
6	30.8% and 43.2%	24.2% and 31.6%	17.4% and 25.9%	11.6% and 22.4%
7	25.9% and 32%	18.5% and 23.8%	13.1% and 19.2%	8.7% and 15.4%
8	20.9% and 24.6%	13.5% and 18.5%	8.9% and 13.8%	5.8% and 10.3%
9	19.75%	13.6%	8.77%	5.85%

$A_2 = \{020(6), 021(7), 022(8), 120(15), 121(16), 122(147, 220(24), 221(25), 222 (26)\}$. When we increase the count of non self-replicating RMTs, first we target set A_1 and then we move to A_2. This work explores all the 3-state 3-neighborhood CAs of sizes 4 to 7 where the corresponding rules maintain 10 (9 from A_0 and 1 from A_1), 11 (9 from A_0 and 2 from A_1), 12 (9 from A_0 and 3 from A_1) and 13 (9 from A_0 and 4 from A_1) non self-replicating RMTs. Table 10 represents the count of the single length cycle CAs based on those non self-replicating RMTs and the proportion of the configurations involved in single length cycles depending on the CA sizes.

Table 10. The relationship among the non self-replicating RMTs, CA sizes and the proportion of single length cycles when the count of non self-replicating RMTs are 10, 11, 12 and 13.

# of selected non -self-replicating RMTs from A_0	# of Rules	# of single length cycle rules	$n = 4$	$n = 5$	$n = 6$	$n = 7$
10	9216	7040	14.8% and 19.7%	8.2% and 13.1%	4.5% and 8.7%	2.4% and 5.8%
11	73728	43424	9.8% and 16%	4.1% and 9.8%	1.6% and 6.5%	0.6% and 4.3%
12	344064	166584	4.9% and 13.5%	1.6% and 8.2%	0.5% and 5%	0.1% and 3.2%
13	1032192	453896	2.5% and 12.3%	0.8% and 7%	0.27% and 4.25%	0.09% and 2.6%

Table 10 shows similar behaviour like Table 9 for establishing the relationship among the non self-replicating RMTs, CA sizes and the proportion of single length cycles. Next, we perform the experimentation on the rules having non self-replicating RMTs up to 25 but here we take samples; we have not explored the entire search space. CAs constructed by the rules with 25 non self-replicating RMTs do not generate any single length cycle. In the next section, we discuss about the role of 3-state single length cycle CAs in pattern classification.

4   Classification

Single length cycle CAs (specially MACA) can be acted as pattern classifier and it has already been validated by the researchers [10,15–18,20]. For designing CA based classifier, we need to select those CAs which can generate a *moderate* number (small percentage of total number of configurations) cycles. In this work, we consider those single length cycle CAs which can maintain less than 20% configurations as single length cycles or point state attractors. Here, we mainly use the single length cycle rules which maintain at least 9 and at most 18 non self-replicating RMTs. If we deal with real datasets, then we need to *encode* the objects into 3-state CA configurations. Every dataset is to be divided for the two phases: *training* and *testing*. In the former phase, we identify the best CA (that gives maximum *accuracy*) for the dataset. In the later phase, we use the CA to detect its power of recognizing and classifying the objects properly which are not trained. Let us describe the process of CA based classification as follows.

1. The dataset is divided into *two* subsets - 70% for *training* and 30% for *testing*.
2. As we deal with real data, we need to encode those data objects into CA configurations. This work only handles those data which possess *quantitativ2.e* attributes. Let us discuss the encoding mechanism:
   – Find the maximum and minimum values of each attribute.
   – Find *range* = maximum value - minimum value.
   – If a value is less than minimum+(1/3)*range, then assign 0 for it.
   – If a value lies between minimum+(1/3)*range, minimum+(2/3)*range, then assign 1 for it.
   – If a value is greater than minimum+(2/3)*range, then assign 2 for it.
   – Next we concatenate all encoded feature values to map each object to a CA configuration, called as *useful* configuration.
3. This encoding scheme may map multiple objects to one configuration.
4. In training phase, a candidate CA (which maintains less than 20% single length cycles) has been selected and we figure out the *useful single length cycles* which can alternatively be called as *useful attractors*. An attractor of a CA is called as *useful* if an useful configuration is attracted to that one i.e., the attractor identifies an object of training dataset of a particular class. Now, an attractor can recognize multiple such training objects from different classes. An attractor is called as the *representative of class i* if it recognizes most objects from class *i* and the count of the recognized objects from that class is called as *objects recognized by the useful attractor*.
5. The training accuracy is calculated as follows:
   $\sum$ objects recognized by the useful attractors/total number of training objects.
6. Among all the candidate CAs, the one giving the best training accuracy is chosen for testing data. In testing an useful attractor can correctly recognize an object only from class *i* if the corresponding attractor is the *representative of class i* in the training phase. The testing efficiency is calculated as follows:
   $\sum$ objects correctly recognized by the useful attractors/total number of testing objects.

This work uses Iris data set from UCI Machine Learning Repository. This data set consists of 150 objects and each data object has *four* quantitative attributes. We code each value of the attributes to one of 0, 1 or 2 by frequency based encoding technique. In this work, we need to use 4-cell CA for the classification of iris data set. Here, we use the single length cycle CAs which maintain 9, 10, 11, 12, 13 and 18 non self-replicating RMTs. We get the best CA 222111222222111211222111122 in training phase with efficiency 97.14286%; the same CA gives the testing efficiency as 97.77778%. This CA maintains 9 non self-replicating RMTs.

5 Conclusion

This paper studies 3-state 3-neighborhood single length cycle cellular automata. From 3^{27}, the subsets of strictly multi length cycle rules and strictly single length cycle rules are identified. This work also presented the efficacy of 3-state 3-neighborhood CAs in pattern classification.

Acknowledgement. This work is carried out as a project in the Summer School on Cellular Automata Technology 2023. The authors are grateful to Dr. Sukanta Das, Associate Professor, Department of Information Technology, Indian Institute of Engineering Science and Technology, Shibpur, for his incessant encouragement, sincere supervision, valuable advice and perfect guidance.

References

1. Adak, S., Naskar, N., Maji, P., Das, S.: On synthesis of non-uniform cellular automata having only point attractors. J. Cell. Autom. **12**(1–2), 81–100 (2016)
2. Chady, M., Poli, R.: Evolution of cellular automaton based associative memories. Technical Report no. CSRP-97-15 (1997)
3. Chakraborty, B., Dalui, M., Sikdar, B.K.: Synthesis of scalable single length cycle, single attractor cellular automata in linear time. Complex Syst. **30**(3), 415–439 (2021)
4. Chaudhuri, P.P., Chowdhury, D.R., Nandi, S., Chatterjee, S.: Additive Cellular Automata – Theory and Applications, vol. 1. IEEE Computer Society Press, USA (1997). ISBN 0-8186-7717-1
5. Chowdhury, D., Gupta, I., Chaudhuri, P.: A low-cost high-capacity associative memory design using cellular automata. IEEE Trans. Comput. **44**(10), 1260–1264 (1995)
6. Das, A.K., Sanyal, A., Palchaudhuri, P.: On characterization of cellular automata with matrix algebra. Inf. Sci. **61**(3), 251–277 (1992)
7. Das, S., Mukherjee, S., Naskar, N., Sikdar, B.K.: Characterization of single cycle ca and its application in pattern classification. In: Electronic Notes in Theoretical Computer Science, vol. 252, pp. 181–203 (2009)
8. Das, S., Naskar, N.N., Mukherjee, S., Dalui, M., Sikdar, B.K.: Characterization of CA rules for SACA targeting detection of faulty nodes in WSN. In: Bandini, S., Manzoni, S., Umeo, H., Vizzari, G. (eds.) ACRI 2010. LNCS, vol. 6350, pp. 300–311. Springer, Heidelberg (2010). https://doi.org/10.1007/978-3-642-15979-4_32

9. Ganguly, N., Maji, P., Sikdar, B., Pal Chaudhuri, P.: Design and characterization of cellular automata based associative memory for pattern recognition. IEEE Trans. Syst. Man Cybern. B Cybern. **34**(1), 672–678 (2004)
10. Ganguly, N.: Cellular automata evolution: theory and applications in pattern recognition and classification. Ph.D. thesis, Bengal Engineering College (a Deemed University), India (2004)
11. Ganguly, N., Das, A., Maji, P., Sikdar, B.K., Pal Chaudhuri, P.: Evolving cellular automata based associative memory for pattern recognition. In: Monien, B., Prasanna, V.K., Vajapeyam, S. (eds.) HiPC 2001. LNCS, vol. 2228, pp. 115–124. Springer, Heidelberg (2001). https://doi.org/10.1007/3-540-45307-5_11
12. Kamilya, S., Adak, S., Das, S., Sikdar, B.K.: SACAS: (non-uniform) cellular automata that converge to a single fixed point. J. Cell. Automata **14**, 27–49 (2019)
13. Kari, J.: Theory of cellular automata: a survey. Theoret. Comput. Sci. **334**(1–3), 3–33 (2005)
14. Maji, P., Shaw, C., Ganguly, N., Sikdar, B.K., Chaudhuri, P.P.: Theory and application of cellular automata for pattern classification. Fund. Inform. **58**, 321–354 (2003)
15. Maji, P.: Cellular Automata Evolution for Pattern Recognition. Ph.D. thesis, Jadavpur University, Kolkata, India (2005)
16. Maji, P., Chaudhuri, P.P.: RBFFCA: a hybrid pattern classifier using radial basis function and fuzzy cellular automata. Fund. Inform. **78**, 369–396 (2007)
17. Naskar, M.B.: Characterization and Synthesis of Non-Uniform Cellular Automata with Point State Attractors. Ph.D. thesis, Indian Institute of Engineering Science and Technology, Shibpur, India (2015)
18. Naskar, N., Adak, S.: Designing multi-class pattern classifier using ca under periodic boundary condition. In: 2018 Fifth International Conference on Emerging Applications of Information Technology (EAIT), pp. 1–4 (2018)
19. Naskar, N., Das, S., Sikdar, B.K.: Characterization of nonlinear cellular automata having only single length cycle attractors. J. Cell. Autom. **7**(5–6), 431–453 (2012)
20. Sethi, B., Roy, S., Das, S.: Asynchronous cellular automata and pattern classification. Complexity **21**(S1), 370–386 (2016)
21. Wolfram, S.: A New Kind of Science. Wolfram Media, Inc. (2002)

Rock Image Classification Using CNN Assisted with Pre-processed Cellular Automata-Based Grain Detected Images

Soumyadeep Paty[1](✉) and Supreeti Kamilya[2](✉)

[1] Department of Mining Engineering, Kazi Nazrul University Asansol, 713340 West Bengal, India
soumyadeep.paty@gmail.com

[2] Department of Computer Science and Engineering, Birla Institute of Technology Mesra, Ranchi, Jharkhand, India
kamilyasupreeti779@gmail.com

Abstract. The mining industry plays a pivotal role in modern society, driving economic growth and sustaining our quality of life. A core challenge within this industry is the accurate identification of rock types, traditionally categorized as igneous, sedimentary, and metamorphic. Conventional methods for this classification, relying on visual assessments, spectroscopy, and chemical analyses, are not only time consuming but also often lack consistency. To address these limitations, this study explores the integration of Convolutional Neural Networks (CNNs) into the mining sector. CNNs, renowned for their image classification capabilities, have seen widespread use in various domains but are underutilized in mining. We focus on classifying rock images with a specific emphasis on granular structures, a promising application previously demonstrated. We introduce Cellular Automata (CAs) to automate grain boundary extraction within rock images. We further investigate the potential of CA-based edge detection coupled with CNNs for rock image classification. Precisely identifying grain boundaries enhances the efficiency of classification techniques, reducing the need for exhaustive iterations. Comparative analysis of the proposed model with two other CNN architectures are shown in the paper. Our proposed model shows superior result compared to the other models. Also, efficiency of CA based model is shown compared to the CNN model without using CA.

Keywords: Rock Identification · Cellular Automata · CNN · Image Classification · Mining · Grain Boundary

1 Introduction

The mining industry holds a central position in the modern era, playing an essential role in sustaining our quality of life and fostering economic growth within nations. One critical aspect of this industry revolves around the efficient

M. Dalui et al. (Eds.): ASCAT 2024, CCIS 2021, pp. 153–167, 2024.
https://doi.org/10.1007/978-3-031-56943-2_12

extraction of minerals, which necessitates the accurate identification of various rock types. These rocks are traditionally classified into three primary categories: igneous, sedimentary, and metamorphic. Traditionally, this classification task has relied heavily on the expertise of skilled professionals [2].

Numerous methods have been developed for rock identification, including visual assessments conducted by the human eye, sophisticated spectroscopic techniques, and detailed chemical analyses [1, 12, 13]. Visual identification, specifically, hinges on the evaluation of various rock attributes, encompassing factors such as color, grain size, luster, and overall structural characteristics. For instance, differentiating between sandstones and shale relies on the observation of their distinct grain sizes, where sandstones exhibit a coarse grain size, while shale typically possesses a finer grain size. Additionally, variations in coloration are employed to distinguish rocks, such as the dark hues of basalts compared to the lighter colors of marble.

However, these conventional rock identification methods are encumbered by inherent limitations, notably their high costs and time intensive nature. Furthermore, achieving consistent and high levels of accuracy through reliance on human expertise remains a persistent challenge. As a result, the scientific community has been diligently working towards a solution that offers a more streamlined, cost-effective, and accurate approach to rock classification, leveraging the capabilities of machine learning techniques.

One of the most well-known facets of deep learning is the Convolutional Neural Network (CNN), which offers a versatile approach for classifying visual images and tackling intricate object recognition tasks. In the contemporary era of computing, CNN has found widespread utility across a multitude of disciplines, spanning social science, brain-computer interfaces, natural language processing, and more. However, the application of convolutional neural networks within the mining industry remains relatively underexplored [3].

While CNN has made significant inroads in diverse domains, its presence in the mining sector has been somewhat limited. For instance, in a study outlined in [4], CNN was harnessed to analyse the granularity of microscopic rock images, shedding light on the potential applications of this technology in the geological realm. There are also different studies used CNN to identify rock images [7, 11, 14]. Additionally, the authors of [10] employed a modified region-based CNN (R-CNN) to detect rocks onsite, achieving a noteworthy accuracy of 80%. This stands as a testament to the unexplored potential of CNN in enhancing efficiency and profitability in mining operations by facilitating accurate rock and mineral identification. In the realm of mining and rock classification, it is widely acknowledged that distinct rock types are composed of minerals with varying grain sizes. Accurate identification of these grain sizes and boundaries is essential, as it has the potential to significantly expedite and enhance the rock image classification process. By precisely delineating grain sizes and boundaries, the classification techniques can operate more efficiently, requiring fewer iterations to arrive at accurate results.

Various image-specific, computer-assisted techniques have been developed to automate the extraction of grain boundaries in digital images. These methods encompass a diverse range of approaches, including microscopic image segmentation through watershed methods, constrained automated seeded region growing algorithms, neural network applications, genetic programming, gradient filtering, and image classification techniques [4,8,14]. Moreover, certain approaches integrate sophisticated edge detection techniques, encompassing methods such as Laplacian edge detection, Gaussian smoothing, Sobel Edge Detection, and various gradient operators.

The primary emphasis of this paper lies in the classification of rock images derived from grain-detected images. The uses Cellular Automata (CAs), a mathematical framework that has found applications in modeling dynamic systems across various domains, including geography, urban dynamics, environmental sciences, and geological applications. However, it's worth noting that the fusion of CA-based edge detection methods and CNN in image processing remains relatively limited. Cellular Automata, in essence, provide mathematical representations of physical systems characterized by discrete space and time. Operating on a standardized grid of cells, these systems engage in local interactions among neighboring cells, defining the state of each cell at various time steps. The state of a given cell at a particular time hinges upon the values of its closest neighbors from the previous time step, governed by a predetermined local transition rule. The amalgamation of outcomes from these localized interactions gives rise to elaborate and dynamic patterns. This renders CAs a powerful instrument for modeling a diverse spectrum of nonlinear and stochastic spatial processes.

This study introduces a model grounded in Cellular Automaton (CA) principles, specifically designed for detecting the edges of rock grains and establishing grain boundaries within images sourced from diverse regions of India. CA shows efficiency in capturing the grain boundary of a rock grain by tracking the borders. The images, capturing various rock types, form a comprehensive dataset representative of the geological diversity across the country. The grain boundary-identified images generated by the CA model serve as pivotal input for our subsequent classification process. The next phase of our methodology seamlessly integrates these grain boundary-enriched images into a CNN model specially crafted for image classification. This CNN-based model is adept at discerning and categorizing the distinct rock types present in the dataset. By capitalizing on the nuanced features extracted through CA-driven edge detection, our CNN model enhances the precision and efficacy of the classification task.

This dual-step approach, combining CA-based grain boundary detection with CNN-based image classification, forms the backbone of our methodology. It not only highlights the versatility of combining traditional image processing techniques with CA based models but also emphasizes the synergy in utilizing the strengths of both approaches for a comprehensive analysis of rock images collected from diverse geographical locations in India.

2   Initial Concepts and Fundamental Terminologies

2.1   Cellular Automation

Let us establish a formal definition for a cellular automaton.

Definition 1. *A cellular automaton (CA) is defined as a quadruple ($\mathscr{L}$,S,$\mathscr{M}$,f), where,*

- *$\mathscr{L} \subseteq \mathbb{Z}^D$ is D-dimensional cellular space, $\vec{v} \in \mathscr{L}$*
- *S is the set of states a cell can assume*
- *$\mathscr{M} = (\vec{v}_1, \vec{v}_2, \cdots, \vec{v}_m)$ is the neighborhood vector of each $\vec{v}$ of the lattice, and $(\vec{v} + \vec{v}_i) \in \mathscr{L}$*
- *$f : S^m \rightarrow S$ is the transition function, called local rule of the CA.*

Definition 2. *(Configuration) A configuration is a mapping that assigns states to the cells $c : \mathscr{L} \rightarrow S$.*

This indicates that $S^{\mathscr{L}}$ is the set of all possible configurations. A CA can also be defined by a function $G : S^{\mathscr{L}} \rightarrow S^{\mathscr{L}}$, which is called as global transition function.

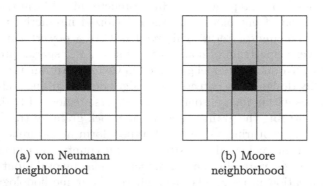

(a) von Neumann   (b) Moore
neighborhood   neighborhood

Fig. 1. Two-dimensional CA neighborhood dependencies in the illustration, the central black cell is the focal point of analysis, with its state influenced by the gray-colored neighboring cells. (Color figure online)

In this research, we focus on finite sized Cellular Automata (CAs) with a dimensionality denoted as D = 2. Typically, the neighborhood of a two-dimensional CA can be defined in one of two main ways. (1) using the von Neumann neighborhood dependency or (2) employing the Moore neighborhood dependency. John von Neumann introduced a specific type of two-dimensional CA characterized by square grid structures.

In the context of two-dimensional Cellular Automata (CAs), the Moore neighborhood adopts a 9-neighborhood dependency, which includes the incorporation of four non-orthogonal cells as neighboring entities. As depicted in

Fig. 1(b), the Moore neighborhood structure has been notably harnessed in the design of renowned CAs like the Game of Life. This particular CA, initially introduced by John Conway and popularized by Martin Gardner, is well-known for its captivating dynamics. In our present study, we also employ the Moore neighborhood configuration in the context of two-dimensional CAs [6].

2.2   Convolutional Neural Networks

Convolutional Neural Networks (CNNs) stand out as a ubiquitous and powerful technology in the realm of deep learning, particularly celebrated for their adeptness in automatic feature engineering within neural networks. Originating from their initial application in the recognition of handwritten digits and letters, as pioneered in [9], CNNs have evolved to become the cornerstone in visual image recognition.

The fundamental operations of a CNN can be dissected into two pivotal phases: feature extraction and classification [5]. In the feature extraction phase, the convolution layer and pooling layer take center stage. These layers work in tandem to automatically discern and emphasize relevant features within the input data. The convolution layer applies filters to the input, capturing spatial hierarchies, while the pooling layer condenses the information, retaining essential details. Conversely, the classification phase unfolds with the fully connected layer, which synthesizes the abstracted features into a comprehensive representation. This layered approach culminates in the classified output images. The architecture of a CNN, illustrated in Fig. 2, serves as a visual road map for the interplay of these components. To elucidate further, the convolution layer and pooling layer collaborate in the intricate task of feature extraction, unraveling the nuances of the input data. Following this, the fully connected layer takes the reins, orchestrating the amalgamation of features and paving the way for the final output classification. The holistic orchestration of these layers encapsulates the essence of a CNN's prowess in deciphering and categorizing complex visual information.

At the core of the CNN architecture lies the convolutional layer, a key player in crafting a detailed feature representation of input images. This layer conducts convolution operations on small blocks of input images through neurons or filters. Non-linear activation functions, such as ReLU or hyperbolic tan, are subsequently applied to enhance the expressive power of the convolutional layer's results.

Following the feature extraction phase, the pooling layer takes center stage, employing operations like max pooling or average pooling to systematically reduce the dimensionality of the feature map. This deliberate reduction enhances the resilience of the output to distortions, fostering a more robust representation of the underlying patterns.

As the CNN journey progresses, the fully connected layer comes into play, adeptly mapping the 2D filters extracted from earlier layers into a one-dimensional array. This layer serves as the final arbiter, leveraging the amalgamated features to classify the input images. It synthesizes the intricacies captured

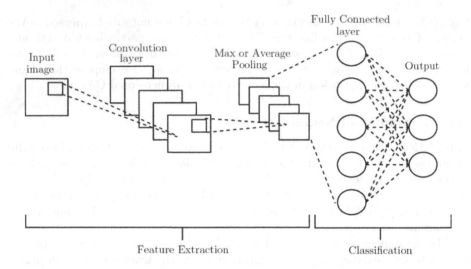

Fig. 2. CNN architecture

in the preceding layers, making informed decisions about the categorization of the visual data.

In essence, the convolution layer spearheads feature representation, the pooling layer bolsters resilience against distortions, and the fully connected layer concludes the process with the classification of images. The orchestrated interplay of these layers characterizes the CNN's ability to distill complex visual information into actionable insights.

3   Description of the Dataset

This study revolves around a dataset comprising images of rocks sourced from various regions across India. The dataset is given in paper [11]. The dataset encompasses three distinct rock types: igneous, sedimentary, and metamorphic. Notably, images of igneous rocks (specifically basalt and granite) and metamorphic rocks were captured from stone quarries in the western part of India, while images of sedimentary rocks (coal, limestone, sandstone, and shale) were obtained from coal mines in the eastern part of the country.

The specific rock types included in the study are as follows: igneous rocks - basalt and granite; sedimentary rocks - coal, limestone, sandstone, and shale; and metamorphic rocks - marble and quartzite. The dataset is robust, comprising 386 images of basalt, 401 images of granite, 380 images of marble, 396 images of quartzite, 369 images of coal, 377 images of limestone, and 390 images of sand-stone and 391 images of shale. In total, there are 787 images of igneous rocks, 1527 images of sedimentary rocks, and 776 images of metamorphic rocks, providing a comprehensive and diverse collection for analysis.

Fig. 3. A Glimpse into Images Featuring Different Rock Types

To offer a visual glimpse into the dataset, a selection of images showcasing different types of rocks is provided in Fig. 3. This dataset serves as a valuable resource for exploring and understanding the distinctive characteristics of igneous, sedimentary, and metamorphic rocks in the context of the Indian geological landscape.

4 Methodology

The methodology employed in this study unfolds in two distinctive phases. In the initial stage, we dedicated efforts to crafting an algorithm tailored for edge detection of rock grains, subsequently creating well-defined grain boundaries. This foundational step lays the groundwork for enhancing the granularity and specificity of our subsequent analyses.

Moving to the second phase, the images enriched with detected grain boundaries become the input for our developed Convolutional Neural Network (CNN) model. The CNN is meticulously designed to classify the images into three overarching classes: igneous, sedimentary, and metamorphic rocks. Furthermore, the model extends its capabilities to an eight-class classification, discerning between specific rock types, namely basalt, granite, marble, quartzite, coal, limestone, sandstone, and shale. This dual-phase approach allows for a nuanced exploration of rock characteristics. The initial edge detection augments the granularity of feature extraction, while the subsequent CNN classification navigates through the intricacies of distinguishing rock types with a fine-tuned precision. Together, these methodological components contribute to a comprehensive understanding of the rock images under scrutiny.

4.1   Grain Boundary Detection Using Cellular Automata

Within the realm of image processing, the foundational technique of edge detection plays a critical role in the nuanced fields of feature detection and extraction. An edge, in this context, is akin to the delineation between two distinct regions, a manifestation of alterations in intensity, color, or texture within a digital image. The primary objective of edge detection is to identify, pixel by pixel, areas where there are abrupt transitions in image brightness levels or significant discontinuities. This process contributes to a clearer understanding of the visual elements within an image [11].

In this investigation, we employ a model for edge detection in digital images, based on CA. Our approach specifically utilizes a two-dimensional CA, where the states within the model represent the varied colors corresponding to each image pixels. This study utilizes Moore's neighborhood CA, emphasizing the collaborative influence of neighboring pixels in the detection process, as the grain boundary can be in any direction. The transition function, a key component of this CA based model, is succinctly expressed as,

$$f(p_1, p_2, p_3, \cdots, p_n) = \begin{cases} 0, & \text{if } |s - s_i| < e \text{ for all } i = \overline{1, n}. \\ p, & \text{otherwise.} \end{cases}$$

where 'e' represents a threshold value critical for discerning regions characterized by pronounced differences in brightness.

The governing transition rule of the model is based on a comparison criterion, evaluating the state of a cell against that of its surrounding cells. It checks the disparity fall below the defined threshold value 'e' or not. If it is below 'e', the state of the central cell in the subsequent time step $(t + 1)$ is set to zero; otherwise, it remains unchanged. This intricate mechanism allows the model to dynamically assess the differences in brightness levels, enabling the identification of regions with significant changes. By delving into the specific interplay between a central cell and its neighbors, this approach refines the process of edge detection, capturing the subtleties of brightness transitions within digital images.

The initial phase of our extraction process involved the application of a 16×16 pixel median filter to mitigate noise originating from various grain features, such as inclusions, cracks, twins, or sub-grains. The algorithm for grain detection in

---

Algorithm 1: Grain Detection Algorithm

Input: Image represented as a matrix of cell values & Threshold value 'e'
Output: Updated image with retained cell states and identified edges
1 Set the threshold value 'e'.
2 **for** *each cell in the image* **do**
3   a. Check the difference of the cell value between the current cell and its neighborhood cells.
4   b. If the difference is less than 'e', then do not change the cell state.

5 **for** *every cell in the image* **do**
6   a. Count the number of connected cells.
7   b. If more than three cells are connected, consider it as an edge.

---

rock images, outlined below, scrutinizes the gray scale color of each cell within a band against that of its neighboring cells.

Significantly, the algorithm's transition rules become active when the difference between colors between the cell and that of its neighboring cells exceeds a predetermined incremental threshold value, denoted as 'e'. This indicates the potential presence of an edge in the cell. In this research, we adopt 'totalistic' rules for the Cellular Automaton (CA) transition, where the $(t+1)$ state of a cell is determined by the sum of the states of all its neighboring cells. This method classifies cells with a state value of 0 as 'non-living' and those with a state value of 1 as 'living'. Moore's neighborhood, encompassing nine possible states ranging from 0 to 9, governs these transitions. Generally, a living cell will transition to a non-living state when surrounded by more than five non-living neighboring cells. Multiple scenarios with different rules and threshold values were explored in our simulations, each yielding optimal grain boundaries through comparison with manually digitized grains using the r^2 assessment criterion.

The final step in our methodology involved the automated creation of a grain boundary database, achieved through a zonal statistics function. This process computed diverse parameters, offering valuable insights into the characteristics of different rocks and contributing to a deeper understanding of their inherent patterns. This comprehensive approach not only refines edge detection but also establishes a robust foundation for understanding the geometric attributes of grains within the dataset.

4.2 Proposed CNN Architecture

This study introduces an innovative CNN architecture designed for the identification of various rock types, employing grain boundary identified images derived from CA based techniques as inputs. Initially, 256×256 grain boundary images of igneous, sedimentary, and metamorphic rocks serve as input for a three-class classification task. Subsequently, the proposed CNN architecture is employed for an 8-class classification, expanding the scope to distinguish be-tween specific rock types within each category.

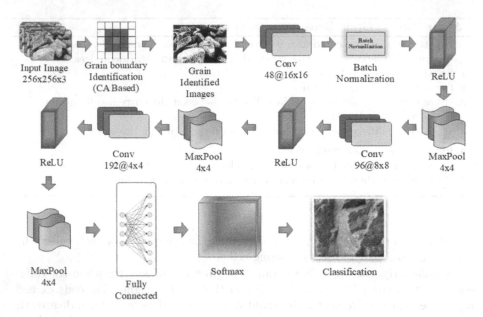

Fig. 4. Proposed Architecture of the CA-assisted CNN Model

To enhance the robustness of the model, image augmentation techniques, including resizing, rotation, and translation, are applied. The augmented images undergo convolutional layer processing, where neurons or filters traverse the input image in different directions, extracting crucial features. For optimized feature extraction and computational efficiency, a gradual reduction in filter size coupled with an increase in the number of filters is implemented across convolution layers.

The proposed architecture, illustrated in Fig. 4, demonstrates an input from the CA based grain boundary identification model to CNN, then in CNN architecture a progressive reduction in filter sizes from 16×16 to 8×8 and 4×4, accompanied by an increment in the number of filters from 48 to 96 and 192. Batch normalization, strategically placed between the convolution layer and ReLU layer, contributes to accelerated training by maintaining consistent speed. The architecture maintains a stride size of 2, and the addition of ReLU layers serves to curb the exponential growth of computations. A 4×4 max pool layer is incorporated after the ReLU layer to downsample feature detection, and a softmax layer is introduced for multi-class classification.

The effectiveness of the proposed CNN architecture is rigorously assessed using a tenfold cross-validation technique, ensuring the model's reliability and generalizability across diverse datasets. This comprehensive approach underscores the model's potential for accurate and robust classification of rock types, offering a valuable contribution to geological image analysis.

5   Results and Discussions

The outcomes derived from the CA model are depicted in Fig. 5, showcasing the visibility of grains under varying threshold values (e = 5, 10, 15, 20). The dark lines in the figure represent the grains identified through the edge detection process. These grain boundary-detected images serve as crucial inputs for the subsequent CNN model.

Fig. 5 demonstrates the alteration in grain-detected images with the variation of the threshold parameter 'e'. Notably, the size and contours of the boundaries undergo noticeable variations with changing threshold values. Larger boundaries manifest when the threshold value is set to a higher magnitude, while smaller boundaries emerge with lower threshold values. This visual representation provides insights into the influence of thresholding on the granularity and characteristics of the identified grain boundaries, essential for understanding and optimizing the subsequent stages of the image analysis process.

The dataset is employed in our proposed CA-CNN based model in this study, where it undergoes classification in two distinct ways. The first classification involves a three-class categorization, distinguishing between igneous, sedimentary, and metamorphic rocks. The second classification expands the scope to an eight-class categorization, covering the subclasses within the aforementioned rock types. The program is executed on MATLAB using an Intel i5 4-core processor with 16 GB RAM.

(a) e=5

(b) e=10

(c) e=15

(d) e=20

Fig. 5. Grain Detected Images for Different Threshold

Performance verification is carried out using standard metrics such as accuracy, recall, precision, and F1-score. These metrics collectively provide a comprehensive evaluation of the model's effectiveness in accurately classifying the rocks based on the grain boundary information extracted through the CA based technique and classified through CNN model. The accuracy, recall, precision, and F1-score are calculated as follows:

$$Accuracy = (TP + TN)/(TP + FP + TN + FN)$$
$$Precision = TP/(TP + FP)$$
$$Recall = TP/(TP + FN)$$
$$F1 - score = 2 \times (precision \times recall)/(precision + recall)$$

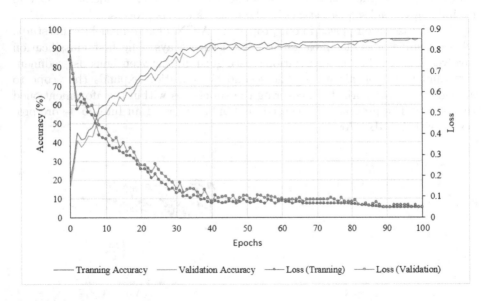

Fig. 6. Graphical Representation of Training and Validation Accuracy & Loss Calculations for the 3-Class Classification

Table 1. A comparative study of our proposed CA-based CNN architecture with the CNN architecture of [11] and GoogleNet

	Proposed CA-CNN Model				On CNN architecture of [11]				GoogleNet			
	Accuracy	Precision	Recall	F1-score	Accuracy	Precision	Recall	F1-score	Accuracy	Precision	Recall	F1-score
3-class classification	93.81	91.22	91.22	91.22	86.47	87.27	90.04	88.21	89.01	84.02	83.92	83.97
8-class classification	97.78	91.18	91.57	91.37	88.93	84.15	86.38	84.98	94.06	77.80	74.29	76.0

Table 2. Comparative analysis of the proposed CA based CNN model and CNN model without CA

	Proposed CA-CNN Combine Model				Proposed CNN model without CA			
	Accuracy	Precision	Recall	F1-score	Accuracy	Precision	Recall	F1-score
3-class classification	93.81	91.22	91.22	91.22	90.84	87.61	87.02	87.32
8-class classification	97.78	91.18	91.57	91.37	94.24	77.47	77.27	77.37

In this context, TP represents true positives, FP denotes false positives, TN signifies true negatives, and FN corresponds to false negatives. The accuracy achieved for the 3-class classification is 94.9%, while for the 8-class classification, it stands at 97.8%. (shown in Figs. 6 and 7).

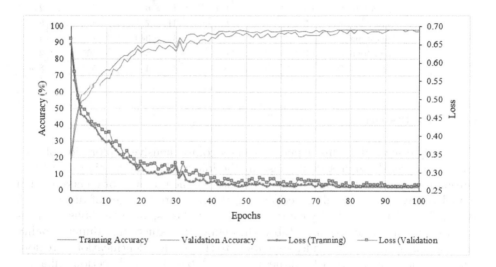

Fig. 7. Graphical Representation of Training and Validation Accuracy & Loss Calculations for the 8-Class Classification

Additionally, we subject the identical dataset to another two CNN architectures, on CNN architecture of [11] and GoogleNet. In this instance, we achieved an accuracy of 93.81% for the 3-class classification and 97.78% for the 8-class classification. It is noteworthy that the proposed architecture outperforms the results obtained with CNN architecture of [11] and GoogleNet. Table 1 further details precision, recall, and F1 score for all cases. The classification is conducted at Epoch 100, employing a tenfold cross-validation methodology, which demonstrates superior performance in categorizing the rock samples. Also, our proposed CA-based CNN model shows better performance compared to the CNN without CA (as shown in Table 2). The confusion matrices for 3-class and 8-class classifications are shown in Table 3 and Table 4.

Table 3. Confusion matrix of 3-class Classification

		Actual Value		
		Igneous Rocks	Sedimentary Rock	Metamorphic Rock
Predicted Value	Igneous Rocks	720	50	27
	Sedimentary Rock	50	1400	50
	Metamorphic Rock	17	77	699

Table 4. Confusion matrix of 8-class Classification

		Actual Value							
		Basalt	Granite	Marble	Quartzite	Coal	Limestone	Sandstone	Shale
Predicted Value	Basalt	329	6	4	3	3	5	4	2
	Granite	7	360	5	6	4	6	2	3
	Marble	8	7	350	6	2	4	3	4
	Quartzite	8	6	4	356	4	6	4	2
	Coal	9	5	4	6	345	7	5	6
	Limestone	8	9	5	5	3	340	2	4
	Sandstone	8	5	2	6	4	5	365	2
	Shale	9	3	6	8	4	4	5	384

6 Conclusion

The study discussed in this paper presents a comprehensive exploration of rock image analysis through a dual-phase methodology. The integration of cellular automaton grain detection and a convolutional neural network classification model proves instrumental in deciphering the intricate features of diverse rock types. The CA model adeptly captures grain boundaries, laying a robust foundation for subsequent analyses. Notably, the deliberate reduction in dimensionality and the nuanced transition rules contribute to the precise identification of regions with significant changes in brightness. This CA-based edge detection enhances the granularity and specificity of the subsequent CNN analyses.

The proposed CNN architecture demonstrates superior performance in classifying rock types compared to the well-known VGG16 model. The strategic design, incorporating gradual filter size reduction and increased filter numbers, ensures optimal feature extraction and computational efficiency. The application of image augmentation techniques further fortifies the model's robustness, showcasing its ability to discern between various rock classes with impressive accuracy.

The dataset, sourced from diverse regions in India, proves to be a valuable resource for geological image analysis. The inclusion of three rock types - igneous, sedimentary, and metamorphic and eight subclasses facilitates a nuanced exploration of distinctive features within the dataset. The results validate the efficacy of the proposed methodology, with high accuracies achieved in both three-class and eight-class classifications. The standard metrics - accuracy, recall, precision, and F1-score underscore the model's reliability in accurately categorizing rocks based on grain boundary information. The comparative analysis with VGG16 further highlights the superiority of the proposed architecture. This study not only advances the understanding of rock image analysis but also showcases the synergy between traditional image processing techniques and state-of-the-art deep learning models. The proposed methodology provides a robust framework for future geological image studies, contributing valuable insights to the field of earth sciences and computational techniques.

References

1. Ashley, G.H., et al.: Classification and nomenclature of rock units. Bull. Geol. Soc. Am. **44**(2), 423–459 (1933)
2. Barton, N., Lien, R., Lunde, J.: Engineering classification of rock masses for the design of tunnel support. Rock Mech. **6**, 189–236 (1974)
3. Chen, J., Yang, T., Zhang, D., Huang, H., Tian, Y.: Deep learning based classification of rock structure of tunnel face. Geosci. Front. **12**(1), 395–404 (2021)
4. Cheng, G., Guo, W.: Rock images classification by using deep convolution neural network. In: Journal of Physics: Conference Series. vol. 887, pp. 012089. IOP Publishing (2017)
5. Elngar, A.A., et al.: Image classification based on CNN: a survey. J. Cybersecur. Inf. Manag. **6**(1), 18–50 (2021)
6. Gardner, M.: The fantastic combinations of Jhon Conway's new solitaire game life. Sci. Am. **223**, 20–123 (1970)
7. Ghorbanzadeh, O., Blaschke, T., Gholamnia, K., Meena, S.R., Tiede, D., Aryal, J.: Evaluation of different machine learning methods and deep-learning convolutional neural networks for landslide detection. Remote Sens.**11**(2), 196 (2019)
8. Gorsevski, P.V., Onasch, C.M., Farver, J.R., Ye, X.: Detecting grain boundaries in deformed rocks using a cellular automata approach. Comput. Geosci. **42**, 136–142 (2012)
9. LeCun, Y., Bottou, L., Bengio, Y., Haffner, P.: Gradient-based learning applied to document recognition. Proc. IEEE **86**(11), 2278–2324 (1998)
10. Liu, X., Wang, H., Jing, H., Shao, A., Wang, L.: Research on intelligent identification of rock types based on faster R-CNN method. IEEE Access **8**, 21804–21812 (2020)
11. Paty, S., Kamilya, S.: Identification of rock images in mining industry: an application of deep learning technique. In: Chakraborty, B., Biswas, A., Chakrabarti, A. (eds.) Advances in Data Science and Computing Technologies. ADSC 2022. LNEE, vol. 1056. Springer, Singapore (2023). https://doi.org/10.1007/978-981-99-3656-4_24
12. Peacock, M.A.: Classification of igneous rock series. J. Geol. **39**(1), 54–67 (1931)
13. Powell, C.M.: A morphological classification of rock cleavage. Tectonophysics **58**(1–2), 21–34 (1979)
14. Zhang, Y., Li, M., Han, S.: Automatic identification and classification in lithology based on deep learning in rock images. Yanshi Xuebao/Acta Petrol. Sinica **34**(2), 333–342 (2018)

A Dynamical Study on Probabilistic Cellular Automata Related to Whale Optimization Algorithm over Time Series Alignment Problems

Tarani Meher[1,2], Anuradha Sahoo[2(✉)], and Sudhakar Sahoo[3]

[1] Department of Mathematics, Government Autonomous College,
Rourkela 769004, India
tysmeher@gmail.com
[2] Department of Mathematics, Siksha 'O' Anusandhan (Deemed to be University),
Bhubaneswar, India
anuradhasahoo@soa.ac.in
[3] Institute of Mathematics and Applications, Bhubaneswar 751029, India
sudhakar.sahoo@gmail.com

Abstract. Alignment of time series of different sampling length are a major topic of research. Many techniques and advance models are formed to tackle the problems. When the alignment is of the multi-objective type, it is more challenging. To answer the problem Cellular Multi-Objective Whale Optimization Algorithm has been developed. This article is based on the effect of the parameters on the performance of the model. We found, some sets of parametric values are enhancing the model, where others have negative effect, demanding more time or high computational cost. Also, we have studied the dynamics of the optimal solution in each time and discussed various cases of changes in Pareto front during the optimization process. The effect of Cellular Automata boundary conditions and neighborhoods on optimization are discussed.

Keywords: Time series alignment · Multi-objective Optimization · Cellular Automata · Pareto front · Whale Optimization Algorithm

1 Introduction

Time series are the data captured at different time from a sensor. Like, heart beats of a patient from ECG machine. Different sensors provides the data in different rates. Alignment of such multi rates time series is a common work in information technology, like, multi-sensors or multi-modal fusion and knowledge discovery [9]. During the alignment of such data, some major challenges arise due to shift or scaled of time and magnitude of the data.

Many authors came with their model to challenge the problems. The "point-wise warping" is a prominent method, where time series are needed to be of

M. Dalui et al. (Eds.): ASCAT 2024, CCIS 2021, pp. 168–181, 2024.
https://doi.org/10.1007/978-3-031-56943-2_13

similar rates to have an alignment with minimum spatial distance. Finding optimal warping path is also a strategy. Many authors used Dynamic programming
and came with algorithm like, Dynamic Time Wraping (DTW). Later, many
variant of (DTW) is developed[15]. It aligns the series with same length, but
when series are of different length it aligns many consecutive points to a single
point of another series[8]. This is a problem, called temporal singularity. Temporal singularity is a common problem in multi-rate time series alignment. Along
with the minimization of spatial distance, minimization of temporal singularity
makes the alignment problem a multi-objective type as minimization of spatial
distance and temporal singularity are conflicting in nature. It is always needed
to find a better algorithm that suit the problem.

Cellular Automata (CA) is used in optimization in various applied fields.
A three dimensional CA is used in [2] for optimizing the block workshop spatial scheduling problem where the dimensions are space, time and import. In
engineering developing circuit or new advance machines CA can be used [5, 6].
In recent studies CA can be used in urban growth simulation and studied the
ecology of a certain place for future use or development [3, 4].

Optimization is an important tool for science and engineering problems.
Depending upon the problems the optimization algorithms are broadly classified
into two types. If only one objective to optimize, single objective optimization
and for more than one objectives, Multi-objectives optimization algorithms are
used. Some optimization algorithms are Genetic Algorithm, Particle Swam Optimization, Whale Optimization Algorithm, etc. In multi-objective optimization
problems, objectives are conflicting in nature. It is not possible to get a single
optimal solution in Multi-objective problems. Therefore, the concept of non-
dominance solutions came. Suppose a minimization problem has n objectives
$f_1, f_2, \cdots, f_n$. If the ith solution dominates jth means,

$$f_k(i) \leq f_k(j), \text{ forall } k = 1, 2, \cdots, n$$

$$f_k(i) < f_k(j), \text{ for atleast one } k \in \{1, 2, \cdots, n\}.$$

If a solution is not dominated by any existing solutions, then it is a non-
dominance solution. The set of all non-dominance solutions form a front in objectives plane, called Pareto front. In Multi-objective optimization, these solutions
are considered as optimal solutions. According to the requirement one is chosen
from the set as final optimal solution.

The Whale Optimization Algorithm (WOA) is based on the hunting techniques of humpback whales[19]. An improved version of this algorithm is proposed in [10] using CA. It is a single objective optimization algorithm considering the whole population as cell space (detail information on CA is discussed in
Sect. 3.4). In [7], a new CA based algorithm WOA-CA is proposed for solving
urban expansion problems. In multi-objective optimization a CA based WOA is
proposed in [1] and named the algorithm as Cell-MOWOA. When an algorithm is
introduced, the detail processes are discussed thoroughly. But in Cell-MOWOA
case, the detail discussion has not been done. It uses many parameters, the effect

of these parameter on the performance is missing. Also why introduction of CA enhances the process has not been discussed.

We have studied this algorithm with different sets of parameters in transition function of CA along with the solution dynamics of the algorithm. The effect of optimization by changing the CA parameters are also discussed.

In this article, Sect. 2 describes the mathematical formulation of Optimization problem, Sect. 3 has a brief discussion on development in WOA, Sect. 4 contains the results and description followed by conclusion in Sect. 5.

2   Mathematical Modeling

Let the length of the low frequency time series is M and that of high frequency time series is L. Each element of low frequency time series (t_i) is aligned with a segment of high frequency series (S_i). Let the length of the segment be $l(S_i)$, for each i from 1 to M, and $\sum_{i=1}^{M} l(S_i) = L$. In each segment S_i, the elements are a_{ij}, for j takes value 1 to $l(S_i)$.

This problem has two objectives, minimization of spatial distance and minimization of temporal singularity. Spatial distance is calculated by the sum of all aligned points' Euclidean distance. Temporal singularity is given by variance of segments length of the high frequency time series. The Mathematical form of the problem is shown below:

$$\min_{\{l(S_1),l(S_2),\cdots,l(S_M)\}} F_1 = \sum_{i=1}^{M} D(S_i, t_i) = \sum_{i=1}^{M} \sum_{j=1}^{l(S_i)} d(a_{ij}, t_i)$$

$$\min_{\{l(S_1),l(S_2),\cdots,l(S_M)\}} F_2 = \sum_{i=1}^{M} (l(S_i) - \bar{l})^2 \qquad (1)$$

such that

$$l(S_i) \geq 1$$

$$l(S_i) \text{ is an integer}$$

where $D(S_i, t_i)$ is the spatial distance between t_i and the segment S_i, $d(a_{ij}, t_i)$ is the Euclidean distance between a_{ij} and t_i and $\bar{l}$ denotes the average of segments length, $\dfrac{L}{M}$. Since the segment lengths are positive integers and each element of the series must be aligned, the constraints are integer values, $l(S_i) \geq 1$.

3   Works on Whale Optimization Algorithm (WOA)

In 2016 Mirjalili and Andrew [19] developed an optimization technique based on hunting style of Humpback whale named as Whale Optimization Algorithm (WOA). It has two phases; (i) Exploration: This is the initial stage of hunting, where whale is moving randomly in the search of food, (ii) Exploitation: It is the later stage where whale attacks the prey using two techniques (a) Encircling and (b) Bobble net. This algorithm can solve many problems efficiently but it shows convergent issue while solving some complex problems. To handle the issue some improvisation to the algorithm have been done like adding a chaotic parameter in the algorithm which is called Chaotic Whale Optimization Algorithm (CWOA)[16], hybridized with differential evolution proposed in [13] called Improved Whale Optimization Algorithm(IWOA). In [7,10], CA based WOA is used to solve urban expansion problems. WOA is used to solve various problems, like construction site lay out problems, power flow, data classification, image retrieval, feature selection [11,14,18] etc. In 2017 [17], Kumawat et al., developed MOWOA. MOWOA is improvised by using a population archive, hybrid with Cellular Automata which is named as Cell-MOWOA [1,12]. The algorithm needs some initial input to start. They are high frequency time series L, low frequency time series M, weights parameters used in cell evolution α, β and γ, spiral control parameter b, population size N, cell space size m and maximum iteration time T.

3.1   Initial Population

The initial population is generated with the knowledge of length of high frequency time series and low frequency time series. In the generation of initial population the objective 'minimization of temporal singularity' is kept in mind. That's why the segment lengths are around the average value $\bar{l} = \dfrac{length(L)}{length(M)}$ with formula

$$X = randi([0.5 \times \bar{l}, 1.5 \times \bar{l}], N, length(M)). \tag{2}$$

It generates N number of random population. Note that, each population is of length equal to length of M. Since it generates the population with random integers, the sum of each population may not equal to length of L. To fix it each population is again checked and made the necessary correction.

3.2   Evaluation of Objective Functions

There are two objectives for each whale population, spatial distance between two time series and temporal singularity. These objective are evaluated by equation (1).

3.3   Non-dominance Value

Multi objective optimization methods used non dominance number as a measure to compare two solutions. Solution with less non-dominance value is a better one. To calculate the non-dominance number of ith whale, set it at 0. For each whale solution $j \neq i$, if $F_1(j) \leq F_1(i)$ and $F_2(j) \leq F_2(i)$, then increase the non-dominance number by 1. In this method non-dominance number of every whales are computed.

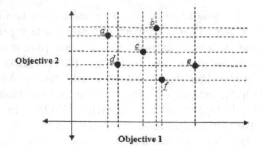

Fig. 1. Sample of whales in objectives plane

Let's consider a set of 6 population in the objectives plane as shown in Fig. 1. To calculate the non-dominance value of each whale, start with whale a. To find the number of whale those dominate whale a draw two lines parallel to the objective axes center at a. It divides the objective plane in 4 quarters. For a minimization problem look inside the lower left quarter along with the drawn lines. The number of whales present inside the region is the non-dominance value of a. Here non-dominance value of a is 0. The non-dominance value of b, c, d, e and f are 3, 1, 0, 2 and 0 respectively.

3.4   Cellular Automata

Cellular automata $CA = (Z_m, St, Nh, f)$ consists of four elements. Cell space Z_m is a space of size m, cell state St is a state of the element in the cell that is used to update the cell in next iteration. Here binary state is considered; '0' for state dead or '1' for alive state. Nh is the set of neighbors of the cells in cell space. Here Moore type neighborhood is considered as shown in Fig. 2. f, the evolutionary function is used for cell updates during iterations.

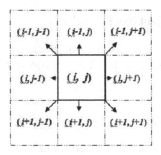

Fig. 2. Moore Neighborhood

– **Cell state evolution**

The cell state depends on multiple factors like, robustness of the element, environmental competitiveness in neighborhood and environmental resource capacity. Here two states '0' and '1' is calculated as follows

$$St^{t+1}(i,j) = \begin{cases} 1, \text{ if } p \leq \alpha \cdot f_{rb}^t(i,j) + \beta \cdot f_{ec}^t(i,j) + \gamma \cdot C_{erc} \\ 0, \text{ otherwise} \end{cases} \tag{3}$$

Here $St^{t+1}(i,j)$ is the state of cell (i,j) for time $t+1$. $f_{rb}^t(i,j)$ and $f_{ec}^t(i,j)$ are robustness and environmental competitiveness of cell (i,j) during time t. C_{erc} is the environmental resource capacity, which is kept fixed as '1'. α, β and γ are weights of f_{rb}^t, f_{ec}^t and C_{erc}, respectively such that $\alpha + \beta + \gamma = 1$.

$$f_{rb}^t(i,j) = \frac{1}{1 + nd^t(i,j)} \tag{4}$$

where $nd^t(i,j)$ is the non-dominance number of the cell (i,j).

To compute environmental competitiveness, first the objective are normalized in the cell space by

$$f_h^t = \frac{F_h^t(i,j)}{\max\limits_{(i',j') \in Z_m} \Gamma_h^t(i',j')}, \qquad h = 1,2. \tag{5}$$

After that crowding distance $cd^t(i,j)$ is evaluated as

$$cd^t(i,j) = -\sum_{(i',j') \in Moore} \frac{f_1^t(i,j) - f_1^t(i',j') + f_2^t(i,j) - f_2^t(i',j')}{ns(i,j)} \tag{6}$$

But for iteration the following steps are used

$$cd^t(i,j) = -\sum_{(i',j') \in Moore} \frac{f_1^t(i,j) - f_1^t(i',j')}{\max F_1} + \frac{f_2^t(i,j) - f_2^t(i',j')}{\max F_2} \tag{7}$$

$$cd^t(i,j) = \frac{cd^t(i,j)}{ns(i,j)} \tag{8}$$

where $ns(i,j)$ is the number of neighbors in the neighborhood of cell (i,j).

$$f_{ec}^t(i,j) = \tanh(cd^t(i,j)) \tag{9}$$

3.5   Update the Whale Population

For whales outside the cell space is updated by usual Whale Optimization Algorithm. Choose a random vector $\overrightarrow{r}$ in [0,1] and a converging sequence $\overrightarrow{a}$ that takes value 2 at initial time and 0 at final time.

$$\overrightarrow{C} = 2\overrightarrow{r}$$

$$\overrightarrow{A} = 2\overrightarrow{a} \cdot \overrightarrow{r} - \overrightarrow{a}$$

choose a random number pn

– (if $pn < 0.5$ and $|\vec{A}| \geq 1$) Choose a random population as prey and calculate

$$\vec{D} = \left| \vec{C} \cdot \vec{X}_{rand}(t) - \vec{X}(t) \right|,$$

Here $|\cdot|$ means absolute value in each tuples.

$$\vec{X}(t+1) = \vec{X}_{rand} - \vec{A} \cdot \vec{D}$$

– (if $pn < 0.5$ and $|\vec{A}| < 1$) prey is the best solution.

$$\vec{D} = \left| \vec{C} \cdot \vec{X}^*(t) - \vec{X}(t) \right|,$$

Here $|\cdot|$ means absolute value in each tuples.

$$\vec{X}(t+1) = \vec{X}^* - \vec{A} \cdot \vec{D}$$

– (if $pn \geq 0.5$) prey is a non-dominance whale solution. Here spiral method is used for update. Choose a random number l to control phase and a fixed spiral shape to control value b.

$$\vec{X}(t+1) = \left| \vec{C} \cdot \vec{X}^*(t) - \vec{X}(t) \right| \cdot e^{bl} \cdot \cos(2\pi l) + \vec{X}(t)$$

The whale alive in the cell space will update. Here a non-dominance solution

$$\vec{X}(t+1) = \vec{X}(t) - \vec{A} \cdot \left| \vec{C} \cdot \vec{X}^*(t) - \vec{X}(t) \right|$$

If the update solution dominates the existing one then solution is replaced by updated one, else no change in whale solution.

3.6   Update of Cell Space

The cell space update is followed by non-dominance sorting of the updated whales. If any outside whale dominates a whale in the cell space then the outside whale will take the place of the worst whale in the space. The detail algorithm is given in Table 1.

4   Results and Discussion

The scheme is a dynamic process and one of the measure part of the scheme is the cell states. The cell states will decide, which whales will update in time $t+1$ from time t. The cell states depend on parameters α, β, γ and probability distribution p. The dot product between a given set of (α, β, γ) with the cell normalized (f_{rb}, f_{ec}, C_{erc}) will provide a value (say state value) for each cell in cell space. When the state value of the cells compare with probability distribution p, the cell state is determined. Then according to cell state, whales are updated. So, it is important to find a better triplet (α, β, γ), for better solution. Also it needs to understand the process to achieve optimal solutions.

Table 1. Cell-MOWOA Algorithm

Input:	L, M, α, β, γ, b, whale population size N, cell space size m, maximum iteration T
1:	Generates the initial whale population $\overrightarrow{X}_i$, $i = 1, 2, \cdots, N$, $t = 0$
2:	Evaluate the objective functions for each population (Fitness)
3:	Compute non dominance value of every whales
4:	Initialization of cell space by non-dominance sorting
5:	**while**$(t < T)$
6:	Choose the prey from non-dominance solution
7:	Calculate state evolution in cell space
8:	Update the whales out side cell space
9:	Update the alive whale in cell space
10:	Update the whales by constraints
11:	Compute the fitness for each population
12:	Compute non-dominance number for updated whales population
13:	Update cell space
14:	$t = t + 1$
15:	**end while**
16:	Find the optimal solution and Pareto front

4.1 Selection of a Better (α, β, γ) Combination

The probability distribution p is chosen randomly, but for our study a fixed probability value is assumed. A whale population is selected and the effect of α, β and γ on the populations is given in Table 2 for a probability value 0.6. Each row of the Table 2 is for a fixed α and β ranges from 0 to $1 - \alpha$. The columns are for fixed β and α varies from 0 to $1 - \beta$. The γ value is calculated by $\gamma = 1 - (\alpha + \beta)$.

Table 2. Percentage of population survive in the cell state

$\alpha \downarrow / \beta \rightarrow$	0.0	0.1	0.2	0.3	0.4	0.5	0.6	0.7	0.8	0.9	1.0
0.0	100	100	100	99.68	48.32	4.56	0.16	0	0	0	0
0.1	100	100	100	48.92	2.32	0.28	0.08	0	0	0	–
0.2	100	100	49.72	1.40	0.44	0.24	0.08	0	0	–	–
0.3	100	52.96	1.08	0.48	0.32	0.24	0.08	0	–	–	–
0.4	100	0.96	0.48	0.40	0.32	0.24	0.08	–	–	–	–
0.5	0.92	0.48	0.48	0.36	0.32	0.24	–	–	–	–	–
0.6	0.48	0.48	0.36	0.36	0.32	–	–	–	–	–	–
0.7	0.48	0.36	0.36	0.36	–	–	–	–	–	–	–
0.8	0.48	0.36	0.36	–	–	–	–	–	–	–	–
0.9	0.36	0.36	–	–	–	–	–	–	–	–	–
1.0	0.36	–	–	–	–	–	–	–	–	–	–

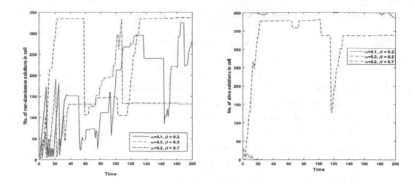

Fig. 3. Non dominance solution and alive cells through out the iterations

The scheme is for optimization and the optimal value is found after the final time. So the final value is more important. The result after maximum time are shown in Fig. 3.

Figure 3 shows a case for small γ value, where very few cells are alive (almost 0) initially (Table 2), and the situation doesn't change much in the later stages. Most of the time there is not a single alive cell in cell space hindering the optimization. Most of the time only the whales outside cell space are updated. Those are not good solutions, so the optimization needs more time to provide a better solution in this conditions.

All the pairs of α and β in Table 2 with non zero entries obtained similar optimal solution but for smaller α and β values (or large γ) make almost all cells in cell space alive (see Fig. 3). This will Update almost all the whale population in next iteration. A high computational option. So a better choice for α, β is the one that update some cells in each time and provide the required solution.

When the probability distribution is considered, all valid values of α, β and γ will provide the optimal solution. But when the γ value is small, the state value will be smaller makes many whales in cell dead. The optimization with very few alive whales demands more time, also for large γ demands maximum computation in each time.

4.2   Pareto Fronts Dynamics

Here the effect of parameters of Cell-MOWOA on optimization is discussed in this section. First, let's consider the effect of CA boundary conditions (see Fig. 4). Three types of boundary conditions are considered. Periodic boundary condition, where next to the last column (row) is taken as the first column (row) and previous to the first column (row) is the last column. Secondly, when the neighbor cells not in the cell space are taken as zero values is called Null boundary condition. Another boundary condition where the neighbor cells not in the cell space are ignored and we named as Wall boundary condition. Figure 4 shows the effect

of CA boundary condition on optimization is very less. In all situations the
algorithm performance are nearly same.

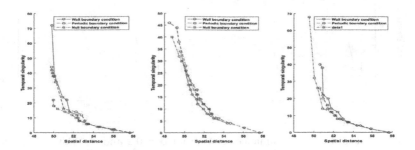

Fig. 4. Effect of boundary conditions for $p = 0.6$. Left: $\alpha = 0.1$ and $\beta - 0.2$, Middle:
$\alpha = 0.2$ and $\beta = 0.3$, Right: $\alpha = 0.2$ and $\beta = 0.7$,

In CA, neighborhood is a very important component and the effect of different
neighborhoods on optimization are shown in Fig. 5. The left and right sub-figure
in Fig. 5 shows no difference in optimization as in these situations the cellular
structure has no measure effect on optimization. But the middle sub-figure says
Moore neighborhood is better. In CA, cells get information from neighbor cells.
Hence the Extended von-Neumann neighborhood gives better result as it's information are received from 8 cells. But the directions are confined to horizontal
and vertical only. When the information provided by horizontal, vertical and
diagonal cells (Moore neighborhood) the performance is further increased. Here
one may think to take extended Moore neighborhood, but then a cell have 24
neighbors. That makes the scheme computationally costly Fig. 5.

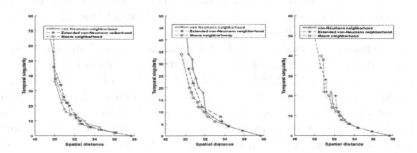

Fig. 5. Effect of neighborhood conditions for $p = 0.6$. Left: $\alpha = 0.1$ and $\beta = 0.2$,
Middle: $\alpha = 0.2$ and $\beta = 0.3$, Right: $\alpha = 0.2$ and $\beta = 0.7$,

Now, consider the non-dominance solutions. Figure 6 shows the number of
non-dominance solution changes in time. These changes are not following any

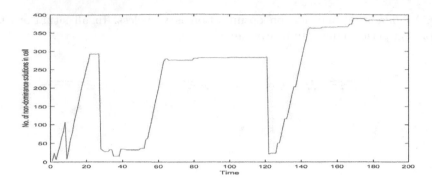

Fig. 6. Non-dominance solution for $\alpha = 0.2$ and $\beta = 0.3$

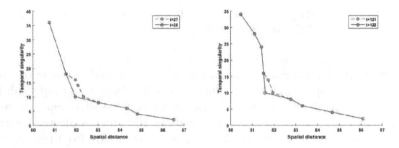

Fig. 7. Pareto front at $t = 27, 28$ and $t = 121, 122$

specific pattern. Some time a sudden drop is observed, also many gradual increments are there. For a few moments the number remain same. In Fig. 6 sudden drop in non-dominance numbers are seen at $t = 27, 107$ and 121. Because at the next time one or more better solutions formed those dominate the previous non-dominance solutions. So, these drops are indicating that the solution are moving towards a better fronts (see Fig. 7).

Now, consider the case for gradual increment of non-dominance solutions over time. In Fig. 6, many times the non-dominance solution number increases, among them two cases $t = 128$ to $t = 131$ and $t = 137$ to $t = 144$ are considered for the study. During the time $t = 128$ to $t = 131$ the Pareto front remain in the same place with no further upgrade. This means new whales solution are joining the front causing an increment in non-dominance solutions number. But this is not the case always. That can be seen in Fig. 9. There are many upgrade in Pareto front at $t = 140$, $t = 141$, $t = 142$ and $t = 143$. Still the non- dominance solutions number increases in each time from $t = 137$ to $t = 144$. That only be possible when the new solutions join the Pareto front are more than the solutions dominated by newly formed solutions.

Similar type of behavior are seen for the case where there is no change in non-dominance solutions number. Here for some time the line remain parallel to time axis in Fig. 6. Observed the segment $t = 100$ to $t = 120$, the non-dominance

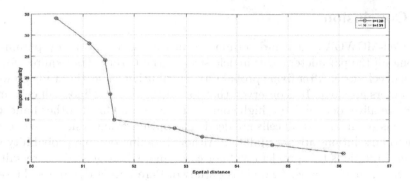

Fig. 8. Pareto front at $t = 128$ and $t = 131$

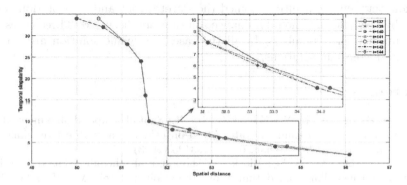

Fig. 9. Pareto front at $t = 137$ to $t = 144$

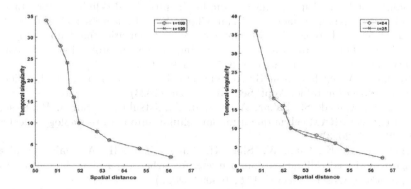

Fig. 10. Pareto front at $t = 100$ and $t = 120$

solutions numbers are unchanged and so as the Pareto front in Fig. 10. But a change is found from $t = 24$ to $t = 25$, though the non-dominance solutions number remain same for $t = 22$ to $t = 26$. So, the solutions dominated by newly formed solution are same to the number added to the Pareto front.

5   Conclusion

The Cell-MOWOA is a non-deterministic process involving many parameters. Among all the parameters, this article studied the parameters required for evolution of cell state. After fixing probability distribution value, the study on weight parameters are done. It is observed that for large γ value, almost all cells in cell space are alive, demanding a high computational cost. On the other hand when γ value is small, almost all cells are dead. The optimization is slow down as most of the time solutions are not updated. Although taking random probability value in every iterations this problem will not cause more damage, still there will be a concern for small γ value. Again in the study of Pareto front dynamics. The effect of CA boundary conditions and neighborhoods are discussed. Optimization with Moore neighborhood gives better results than von-Neumann neighborhood. The relation between the solutions joined the Pareto front and the solutions dominated by newly formed non-dominance solution are discussed. Three cases arise by comparing their numbers. In all the cases the whale solution are updated towards the optimal solutions.

References

1. Liang, B., Han, S., Li, W., Huang, G., He, R.: Spatial-temporal alignment of time series with different sampling rates based on cellular multi-objective whale optimization. Inf. Process. Manag. **60**(1), 103123 (2023)
2. Chen, Y., Lin, X., Yi, W.: Block workshop spatial scheduling based on cellular automata modelling and optimization. IET Collab. Intell. Manuf. **5**(1), e12075 (2023)
3. Zhang, P., et al.: Exploring the response of ecosystem service value to land use changes under multiple scenarios coupling a mixed-cell cellular automata model and system dynamics model in Xi'an. China. Ecol. Indicators **147**, 110009 (2023)
4. Ou, D., et al.: Ecological spatial intensive use optimization modeling with framework of cellular automata for coordinating ecological protection and economic development. Sci. Total Environ. **857**, 159319 (2023)
5. Wanna, P., Wongthanavasu, S.: An efficient cellular automata-based classifier with variance decision table. Appl. Sci. **13**(7), 4346 (2023)
6. Bhat, S., Ahmed, S., Bahar, A., Wahid, A., Otsuki, A., Singh, P.: Design of cost-efficient SRAM cell in quantum dot cellular automata technology. Electronics **12**(2), 367 (2023)
7. Ding, Y., Ao, K., Qiao, W., Shao, H., Yang, Y., Li, H.: A whale optimization algorithm-based cellular automata model for urban expansion simulation. Int. J. Appl. Earth Observ. Geoinf. **115**, 103093 (2022)
8. Jiang, Y., et al.: EventDTW: an improved dynamic time warping algorithm for aligning biomedical signals of nonuniform sampling frequencies. Sensors **20**(9), 2700 (2020)
9. Chen, H., Cai, M., Xiong, C.: Research on human travel correlation for urban transport planning based on multisource data. Sensors **21**(1), 195 (2020)
10. Gao, Y., Tao, Z., Wu, J., Qian, C., Zhou, H., Yang, Y.: Improved whale optimization algorithm via cellular automata. In: 2020 IEEE International Conference on Progress in Informatics and Computing (PIC), pp. 34–39. IEEE (2020). https://doi.org/10.1109/PIC50277.2020.9350796

11. Rana, N., Latiff, M., Abdulhamid, S., Chiroma, H.: Whale optimization algorithm: a systematic review of contemporary applications, modifications and developments. Neural Comput. Appl. **32**, 16245–16277 (2020)
12. Got, A., Moussaoui, A., Zouache, D.: A guided population archive whale optimization algorithm for solving multiobjective optimization problems. Expert Syst. Appl. **141**, 112972 (2020)
13. Mostafa, B., Yazdani, S.: IWOA: an improved whale optimization algorithm for optimization problems. J. Comput. Des. Eng. **6**(3), 243–259 (2019)
14. Gharehchopogh, F., Gholizadeh, H.: A comprehensive survey: whale optimization algorithm and its applications. Swarm Evol. Comput. **48**, 1–24 (2019)
15. Folgado, D., Barandas, M., Matias, R., Martins, R., Carvalho, M., Gamboa, H.: Time alignment measurement for time series. Pattern Recogn. **81**, 268–279 (2018)
16. Kaur, G., Arora, S.: Chaotic whale optimization algorithm. J. Comput. Des. Eng. **5**(3), 275–284 (2018)
17. Kumawat, I., Nanda, S., Maddila, R.: Multi-objective whale optimization. In: Tencon 2017-2017 IEEE Region 10 Conference, pp. 2747–2752. IEEE (2017)
18. Mafarja, M., Mirjalili, S.: Hybrid whale optimization algorithm with simulated annealing for feature selection. Neurocomputing **260**, 302–312 (2017)
19. Mirjalili, S., Lewis, A.: The whale optimization algorithm. Adv. Eng. Softw. **95**, 51–67 (2016)

Synthesis of Fault-Tolerant QCA Logic Circuit Using Cellular Automata

Amit Kumar Pramanik[1,3](✉), Jayanta Pal[2], and Bibhash Sen[3]

[1] Department of CSE, Dumka Engineering College, Dumka, Jharkhand, India
[2] Department of Information Technology, Tripura University, Suryamaninagar, Tripura, India
jayantapal@tripurauniv.ac.in
[3] Department of CSE, National Institute of Technology, Durgapur, WB, India
akp.18cs1509@phd.nitdgp.ac.in , bsen.cse@nitdgp.ac.in

Abstract. The CMOS technology faces significant obstacles in digital logic design because of its reduced device density and increased power consumption. Quantum-dot-cellular automata (QCA) has arisen as a promising alternative to address these limitations. The basic unit for constructing logic circuits in QCA is the QCA cell, comprising four quantum dots with two electrons. In QCA, the fundamental building blocks are the majority gate and inverters. Clocking has a significant contribution in guaranteeing proper synchronization and data propagation within a circuit. Additionally, regular clocking addresses manufacturing challenges in the nanoscale domain and promotes scalability. Nonetheless, defects continue to be a worrisome issue in the realization of nanoscale circuits. In this regard, this work demonstrates the synthesis of fault-tolerant QCA circuits utilizing Cellular Automata (CA) rules. The Hardware Description Language (HDLQ) and QCADesigner simulators serve this particular function.

Keywords: QCA · Fault Tolerance · CA · HDL · QCADesigner

1 Introduction

Since CMOS technology dissipates more power and has a lower device density, it is currently facing significant obstacles. One of the most viable solutions is quantum-dot-cellular automata (QCA) [1] to address these issues. The fundamental element in QCA consists of a cell that includes four quantum dots. Two electrons are positioned in the diagonal quantum dots because of the Coulombic interaction force among the cells. The majority gate and inverter are the two fundamental logic in QCA that can be utilized to design any QCA circuits. The majority gate can be utilized to design conventional AND and OR gates by setting constant values to any one of the three inputs.

The role of clocking in QCA circuits is crucial as it facilitates the movement of electrons between quantum dots. QCA clocking typically involves four clock

M. Dalui et al. (Eds.): ASCAT 2024, CCIS 2021, pp. 182–191, 2024.
https://doi.org/10.1007/978-3-031-56943-2_14

zones: switch, hold, release, and relax. Regular clocking is instrumental in the successful implementation of QCA logic circuits, ensuring their scalability and reliability. Various regular clocking schemes have been documented in references [2–4]. The USE [2] clocking scheme enables bidirectional data flow, while the RES [3] and Zig-Zag [4] clocking schemes support three-way data flow.

The fundamental concept of circuit design is the correct positioning of QCA cells to obtain the desired output. However, a critical concern at the nanoscale level is the presence of defects, and the correctness of a circuit relies on its ability to withstand faults. Typically, defects can manifest during the cell deposition phase, the synthesis phase, or both, with the highest likelihood of defects occurring during cell deposition. Common faults in QCA circuits include cell displacement, omission, the addition of extra cells, and cell misalignment [5]. These defects can arise due to thermodynamic effects or minor energy variations between excited and ground states within the circuit. As a result, a thorough investigation of fault tolerance is essential to ensure the feasibility of any circuit.

Cellular automata (CA) are dynamic in nature and exhibit intricate global tendencies arising from basic regional interactions and calculations. In the 1950 s, the idea of CA was first conceptualized by von Neumann. In the 90 s, Wolfram [6] demonstrated that the family of basic 1-dimensional CA might simulate complicated characteristics [7, 8]. The suggested CA structure was conceptualized as a discrete lattice consisting of cells with two states each, featuring a 3-neighborhood dependence, which included self, left, and right neighbors. One particular category of Wolfram's 1-dimensional CA with 3-neighborhood, referred to as linear CA, garnered significant interest [9]. Since its conceptualization, CA has drawn the interest of numerous researchers from a variety of areas and backgrounds to represent diverse physical, natural, and real-world phenomena. Traditionally, CAs are considered uniform. Nonetheless, non-uniformity has been found in various aspects, including the modified structure, lattice arrangement, neighborhood dependency, and local rules.

Motivated by the above aspects, a synthesis mechanism is identified using the CA to design QCA circuits efficiently with respect to fault-tolerance capability. In this context, the USE clocking scheme is employed to realize the circuits. The primary elements of the article are outlined as follows:

- An efficient synthesis of QCA circuits using CA rules concerning fault-tolerant capability.
- The efficacy of the presented structures is analyzed with respect to the ability to tolerate faults.
- In this aspect, QCADesigner and HDLQ tools are employed.

The arrangement of the article is as follows: a short overview of the fundamentals of QCA, clocking, various faults, and CA is given in Sect. 2. Section 3 offers an in-depth exploration of the realization of QCA circuits and fault-tolerant analysis. Section 4 assesses the design's efficiency and conducts an evaluation of performance. The conclusion has been drawn in Sect. 5.

2 Fundamentals and Prior Work

The smallest element in QCA is a cell, which is made up of four quantum dots and two electrons, as presented in Fig. 1a. The intra-cell coulombic force in a QCA cell determines the orientation of the electrons. In a QCA cell, the electrons preferentially move along the maximum diagonal distance. There are two conceivable configurations of quantum dots in a cell: symmetric and regular. Apparently, two polarization states are possible for the QCA cell (denoted as P), whereas +1 represents a logic 1, while −1 corresponds to a logic value of 0, represented in Fig. 1b. The basic gates in QCA are the majority gate and inverter. The majority gate has one output and three inputs, as illustrated in Fig. 1c.

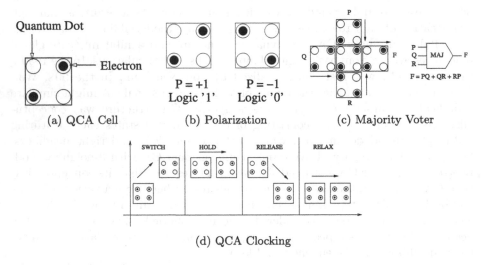

(a) QCA Cell   (b) Polarization   (c) Majority Voter

(d) QCA Clocking

Fig. 1. Fundamental Elements of QCA

QCA clocking [10] is employed to ensure the synchronization of data propagation within the circuit and operates in four distinct phases, as depicted in Fig. 1d. QCA clocking is employed to manipulate the tunneling barriers between quantum dots, enabling the transportation of electrons across them. The idea of clocking was initially proposed by [11], and it has since become a fundamental aspect of QCA circuit design. Nowadays, there are several popular clocking schemes available for implementing QCA circuits, one of which is the USE Clocking Scheme [2].

QCA exhibits wide fault rates due to different cell placement defects [12]. These faults can take place during the deposition phase or fabrication phase of the QCA circuit. Various types of defects are possible. A cell is incorrectly positioned compared to its intended position in a cell misalignment defect. Another kind of fault is cell displacement, where a cell is dislocated from its correct location. On the other hand, the extra cell defect involves the addition of extra cell(s) compared to the actual arrangement of cells. A cell missing defect occurs when a cell disappears from its actual location. Rotation fault occurs when a cell is placed with the opposite orientation (45° in place of 90° or the opposite way).

A Cellular Automaton (CA) comprises multiple cells arranged in a lattice structure. It operates within discrete space and time, resembling an independent finite state machine. Every cell retains a discrete variable at a time, 'i', which denotes the cell's current state (CS). The succeeding state (ss) of a cell at $(i+1)$ is influenced by its own state and the states of its adjacent cells at a time 'i'. In this study, the focus has been given to a 3-neighborhood CA, which considers interactions with the self, left, and right adjacent. In this configuration, each CA cell has two possible states, 0 or 1, and the subsequent state of the jth CA cell is determined as Eq. 1.

$$C_j^{i+1} = f_j(C_{j-1}^i, C_j^i, C_{j+1}^i) \qquad (1)$$

In Eq. 1, f_j is the succeeding state function, and C_{j-1}^i, C_j^i, and C_{j+1}^i are the current states of the jth cell's left adjacent, own, and right adjacent at time i. The cell states denoted as $C^i = (C_1^i, C_2^i ... C_n^i)$ at time 'i', represent the current state of the CA. Hence, the succeeding state of an nth cell CA is calculated as Eq. 2.

$$C^{i+1} = (f_1(C_0^i, C_1^i, C_2^i), f_2(C_1^i, C_2^i, C_3^i), ..., f_n(C_{n-1}^i, C_n^i, C_{n+1}^i)) \qquad (2)$$

The jth CA cell's succeeding state function can be displayed in tabular form by applying specific rules outlined in Table 1. In a two-state and three-neighborhood CA, there can be a total of $2^8 = 256$ different rules [6]. The 1st row in the Table 1 represents the probable ($2^3=8$) current states of $(j-1)^{th}$, $(j)^{th}$ and $(j+1)^{th}$ at time 'i'. The next rows display the jth cell's subsequent states at $(i+1)$ for various combinations of its neighbors' current states utilizing few specific rules.

3   Realization of Circuits Using CA Rules and Their Fault-Tolerant Analysis

This article demonstrates a realization mechanism to design efficient QCA circuits from the sum of product expressions utilizing various CA rules. In this regard, six Cellular Automata (CA) rules $(15, 51, 85, 170, 204, 240)$ are identified from 256 rules. The subsequent state for a particular CA cell is tabulated in Table 1 using the above-mentioned CA rules. Three different boolean functions are used to illustrate the presented mechanism. At the same time, the efficacy of the presented mechanism is analyzed with respect to fault-tolerant capability. HDLQ and QCADeisgner tools are utilized for this purpose.

Table 1. Succeeding Sate for j^{th} CA Cell using CA Rules

PS	111	110	101	100	011	010	001	000	Rule
SS	0	0	0	0	1	1	1	1	15
SS	0	0	1	1	0	0	1	1	51
SS	0	1	0	1	0	1	0	1	85
SS	1	0	1	0	1	0	1	0	170
SS	1	1	0	0	1	1	0	0	204
SS	1	1	1	1	0	0	0	0	240

The HDLQ simulator is used to analyze all probable faults associated with a particular cell [13]. This simulator encompasses components such as inverters, majority voters, wire crossings, L-shaped structures, and fan-outs, along with the ability to introduce faults. In this aspect, the HDLQ models for all circuits are represented in Table 2 through Table 4. In these models, symbols such as "L", "F", "M", "I", and "W" indicates an L-shaped, Fan-out, Majority gate, Inverter, and Wire Crossing, respectively. These diagrams are rigorously evaluated [14] for different types of defects to assess their fault tolerance and overall performance. On the other hand, QCADeisgner is used to check the ability of a QCA circuit to tolerate faults by removing one cell (other than input, output, and constant polarization cells) at a time. Therefore, the QCA layout for all the circuits is represented in Table 2 through Table 4. In these layouts, 'C' denotes those cells which do not affect the output of the circuits. On the other hand, 'F' denotes those cells which influence the output of the circuits.

3.1   Case 1

A function O_1 (Eq. 3) is considered in this section. The sum of product expression can be represented in the form of majority expression as mentioned in Eq. 4. Now, the Eq. 4 can be rearranged by using CA rule 204 and boolean algebra in the form of Eq. 5.

$$O_1 = \Sigma(m_1, m_3, m_4, m_5, m_7, m_{12}, m_{13}, m_{15}) \tag{3}$$

$$M_1 = M((M(M(A', D, 0), M(B, D, 0), 1), M(B, C', 0), 1)) \tag{4}$$

$$R_1 = M(M(C', 0, B), D, M(1, A', B)) \tag{5}$$

3.1.1   Fault-Tolerant Analysis

The efficiency of the previously mentioned technique (Sect. 3.1) is demonstrated by comparing the fault-tolerant capability of both the circuits, presented in Eq. 4 and Eq. 5. In this aspect, the HDLQ model and the QCA layout for these circuits are presented in Table 2.

Table 2. HDLQ Models and QCA Layouts for Case 1

Equation	HDLQ Model	QCA Layout
4 (Before applying Rule)		
5 (After applying Rule)		

3.2   Case 2

Another function O_2 is considered in this case, and it is presented in Eq. 6. The equivalent majority expression for Eq. 6 is presented in Eq. 7. Firstly, Rule 240 is applied on the Eq. 7 and then Rule 170 is applied to the resultant expression. Finally, boolean algebra is used to produce the rearranged expression displayed in Eq. 8.

$$O_2 = \Sigma(m_2, m_4, m_6) \tag{6}$$

$$M_2 = (M(M(A, C', 0), M(B, C', 0), 1)) \tag{7}$$

$$R_2 = M(M(1, B, A), 0, C') \tag{8}$$

The application of CA rules is demonstrated in the diagram as represented in Fig. 2.

3.2.1   Fault-Tolerant Analysis To investigate the efficiency of the mechanism, the fault-tolerant ability of the final circuit is compared with the original circuit (Eq. 7) by injecting faults at different positions of the circuit and removing one cell at a particular time. Therefore, the HDLQ diagrams and QCA layout of both circuits are portrayed in Table 3.

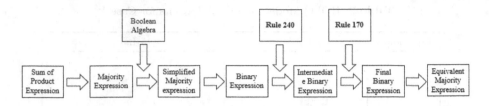

Fig. 2. Application of CA Rules

Table 3. HDLQ Models and QCA Layouts for Case 2

Equation	HDLQ Model	QCA Layout
7 (Before applying Rules)		
8 (After applying Rules)		

3.3 Case 3

The function O_3 is realized in this case using CA rules. Hence, an equivalent majority expression (M_3) is obtained for the O_3. After that, boolean algebra, Rule 85, and Rule 51 are applied to the majority expression. Finally, Rule 15 and, again, Rule 51 are applied to the resultant function to obtain the final expression R_3 (Eq. 11).

$$O_3 = \Sigma(m_3, m_6, m_7, m_{11}, m_{13}, m_{14}, m_{15}) \tag{9}$$

$$M_3 = M(M(M(B, D, 0), A, 0), M(M(B, D, 1)C, 0), 1) \tag{10}$$

$$R_3 = M(C, M(B, C, D), M(A, D, 0)) \tag{11}$$

3.3.1 Fault-Tolerant Analysis

Similar to the other cases, the capability of the function R_3 to tolerate faults is investigated using HDLQ and QCADesigner compared to M_3. In this regard, diagrams and layouts are displayed in Table 4.

Table 4. HDLQ Models and QCA Layouts for Case 3

Equation	HDLQ Model	QCA Layout
4 (Before applying Rule)		
5 (After applying Rule)		

4 Results and Discussion

The outcomes of the Hardware Description Language simulator (HDLQ) for the above three cases are tabulated in Table 5. According to the table, the ability to tolerate the faults is improved in each case after applying the CA rules. In the same way, the outcomes of the QCADesigner for all the cases are depicted in Table 6. It is also observed from the last table that these CA rules can be used to realize fault-tolerant QCA circuits, except for Case 2. In case 2, it is noticed that after the application of the CA rules, the circuit is extremely minimized. As a result, all the cells are very important to provide correct output. Therefore, the fault-tolerant ability deteriorates when QCADesigner is used for this function.

Table 5. The Results of HDLQ Simulator

Study	Fault-Tolerant Capability (in %)	
	Before applying Rule	After applying Rule
Case 1	75.43	85.5
Case 2	57.5	68.63
Case 3	75.18	82.5

Table 6. The Outcomes of QCADesigner

Study	Before applying Rule			After applying Rule		
	Total Cell	Count of 'C' Cell	Fault-Tolerant (in %)	Total Cell	Count of 'C' Cell	Fault-Tolerant (in %)
Case 1	147	108	73	93	71	76
Case 2	96	63	65	22	8	36
Case 3	79	42	53	118	91	77

5   Conclusion

The rules of Cellular Automata (CA) represent significant characteristics both visually and statistically for CA. This paper demonstrates a CA rules-based mechanism for the synthesis of the efficient QCA circuit from the Sum of product expression. The effectiveness of the presented mechanism is analyzed with respect to the ability to tolerate the defects. In this regard, HDLQ is used to model the QCA circuit for fault tolerance analysis, whereas QCADesigner is used to investigate single-cell omission. According to the analysis, CA rules have a significant impact on the design of fault-tolerant QCA circuits. In the best case, rules-based synthesis achieves 45% higher fault-tolerant capability. Therefore, CA rules can be employed for the efficient synthesis of QCA circuits.

References

1. Karim, F., Walus, K.: Efficient simulation of correlated dynamics in quantum-dot cellular automata (QCA). IEEE Trans. Nanotechnol. **13**(2), 294–307 (2014)
2. Campos, C.A.T., Marciano, A.L., Neto, O.P.V., Torres, F.S.: Use: a universal, scalable, and efficient clocking scheme for QCA. IEEE Trans. Comput. Aided Des. Integr. Circuits Syst. **35**(3), 513–517 (2015)
3. Goswami, M., Mondal, A., Mahalat, M.H., Sen, B., Sikdar, B.K.: An efficient clocking scheme for quantum-dot cellular automata. Int. J. Electron. Lett. **8**(1), 83–96 (2020)
4. Pal, Jayanta, Pramanik, Amit Kumar, Sharma, Jyotirmoy Sil, Saha, Apu Kumar, Sen, Bibhash: An efficient, scalable, regular clocking scheme based on quantum dot cellular automata. Analog Integr. Circ. Sig. Process. **107**(3), 659–670 (2021). https://doi.org/10.1007/s10470-020-01760-4
5. Pramanik, A.K., Pal, J., Mohit, K., Goswami, M., Sen, B.: Cost-efficient method for inverter reduction and proper placement in quantum-dot cellular automata. Int. J. Electron., 1–34 (2022). https://doi.org/10.1080/00207217.2022.2145503
6. Wolfram, S.: Cellular Automata and Complexity. Collected Papers (1994)
7. Martin, O., Odlyzko, A.M., Wolfram, S.: Algebraic properties of cellular automata. Commun. Math. Phys. **93**, 219–258 (1984). https://doi.org/10.1007/BF01223745
8. Wolfram, S.: Statistical mechanics of cellular automata. Rev. Mod. Phys. **55**, 601–644 (1983). https://doi.org/10.1103/RevModPhys.55.601

9. Maji, P., Shaw, C., Ganguly, N., Sikdar, B.K., Chaudhuri, P.P.: Theory and application of cellular automata for pattern classification. Fundam. Inf. **58**(3–4), 321–354 (2003)
10. Pramanik, A.K., Pal, J., Sen, B.: Impact of genetic algorithm on low power QCA logic circuit with regular clocking (2022)
11. Lent, C.S., Tougaw, P.D.: A device architecture for computing with quantum dots. Proc. IEEE **85**(4), 541–557 (1997)
12. Tahoori, M., Huang, J., Momenzadeh, M., Lombardi, F.: Testing of quantum cellular automata. IEEE Trans. Nanotechnol. **3**, 432–442 (2005)
13. Fijany, A., Toomarian, B.N.: New design for quantum dots cellular automata to obtain fault tolerant logic gates. J. Nanopart. Res. **3**(1), 27–37 (2001)
14. Pramanik, A.K., Pal, J., Priyadarshini, S., Sen, B.: Fault tolerant QCA logic circuit using genetic algorithm under regular clocking scheme. J. Cell. Autom. **1**(2), 25–45 (2023)

A Note on α-Asynchronous Life-Like Cellular Automata

Souvik Roy[1(✉)], Subrata Paul[2], and Sumit Adak[3]

[1] Ahmedabad University, Ahmedabad, Gujarat, India
souvik.roy@ahduni.edu.in,svkr89@gmail.com
[2] Indian Institute of Engineering Science and Technology, Shibpur, India
[3] Technical University of Denmark,Kongens Lyngby, Denmark
suad@dtu.dk

Abstract. This note shows the dynamics of `Life`-like cellular automata under α-asynchronous perturbation where each cell is updated with α probability. Here, we explore the possibility of phase transition dynamics during evolution of both low and high density `Life`-like games. Hereafter, we compare Game of Life (`Life`) and `Life`-like games with the effect of perturbation. This study also displays a beautiful gallery (natural patterns) of extended `Life` games with the effect of perturbation. Finally, we explore random games and their connection with second-order phase transition and first-order irreversible phase transition.

Keywords: Cellular Automata · Game of Life · Life-like games · α-asynchronism · Phase transition

1 Introduction

"Since the study of life began, many have asked: is it unique in the universe, or are there other interesting forms of life elsewhere? Before we can answer that question, we should ask others: What makes life special? If we happen across another system with life-like behavior, how would we be able to recognize it?" – David Eppstein [14].

In 1970, the late english mathematician John Horton Conway have introduced the most popular Game of `Life` (or, we can simply call `Life`) cellular automata which can able to generate many complex phenomenon and computational universality [3,6,12,17]. Hereafter, the `Life` research community have continuously asked the fundamental questions – what makes `Life` too special? In this direction, the early work of Torre and Mártin [13] have displayed the gallery of extended `Life` games (rules) and their connection with patterns of nature, like 'labyrinths', 'island with borders', 'ferromagnetic domains', 'solid-liquid mixture'. On the other hand, `Life` community have identified many `Life`-like games which shows dynamics most similar to `Life` [2,14,18–20]. In terms of density, some `Life`-like games are associated with low-density (like, `Eight`

M. Dalui et al. (Eds.): ASCAT 2024, CCIS 2021, pp. 192–203, 2024.
https://doi.org/10.1007/978-3-031-56943-2_15

Life, High Life, Honey Life, Pedestrian Life etc.) and some games are with high-density (like Dry Life, Drigh Life etc.). In a recent work, Peña and Sayama [26] have explored the quantitative comparison of Life-like games using the notion of information gain complexity or conditional entropy. According to [26], Life shows a consistent amount of complexity throughout its evolution in comparison with low and high density Life-like games. A good survey about Life-like games and their capability of generating gliders, bombers, rakes, small oscillators, still life, spaceships and other patters can found in [14].

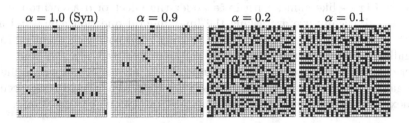

Fig. 1. Space-time diagrams of Life for changing value of α after transient time steps.

On the other hand, the question of randomness (or noise) is fundamental to understand the laws of real life [32]. In this direction, cellular automata research community have also explored the relationship between Life rules and real life (natural systems). Specifically, the fundamental question – 'how real the Life rules are?'. On other words, the subject of study is to understand the effect of perturbation (or noise) in Life [1, 10, 15, 16, 24, 27, 30]. In the first attempt, Schulman and Seiden [30] have introduced the noise with the notion of stochastic component 'temperature' in Life. However, in the construction of [30], the transition probabilities are dependent on the global density, i.e. the rules are not local. In the same direction, Adachi, Peper and Lee [1] have observed a phase transition in Life (from '$1/f$' phase to 'Lorentzian spectrum' phase) for 'temperature' based continuous perturbation. Under fully asynchronous updating scheme, Bersini and Detours [7] have also observed the phase transition in Life. Fatès [15, 16] have asked the important question – 'does Life resist asynchrony?' many times. In our early work [27], we have observed the continuous and abrupt change in phase during evolution of Life with probabilistic loss of information and delay (during information sharing between two neighbouring cells) perturbation respectively. In this context, following are the other notable contributions – γ-asynchronous Life [15, 16]; Life with short-term memory [5]; quantum Life [8, 25]; probabilistic Life [4]; temporally stochastic Life [22, 29]. Moreover, during the evolution of stochastic Life rules, Monetti and Albano [23, 24] have observed first-order irreversible phase transition from steady state (non-zero density) to extinct phase (zero density). A good survey about Life and perturbation can found in [21].

In particular, in this present article, our topic of interest is dynamics of Life and Life-like games under α-asynchronous perturbation. Fatès and Michel

[15, 16] have identified a phase transition of Life from an 'inactive sparse' phase to 'labyrinth' phase for changing value of α where each cell is updated with α probability. For evidence, Fig. 1 depicts the dynamics of Life after transient (1000 time steps) steps for (a) synchronous, i.e. $\alpha = 1.0$; (b) $\alpha = 0.9$; (c) $\alpha = 0.2$; (d) $\alpha = 0.1$ environment. Note that, Life shows labyrinth phase for $\alpha = 0.1, 0.2$ in Fig. 1. Following the literature, the current study explores the dynamics of low-density (Eight Life, High Life, Honey Life, Pedestrian Life, Flock Life, LowDeath Life etc.) and high-density (Dry Life, Drigh Life etc.) Life - like games under α-asynchronous perturbation. Hereafter, we compare Life - like games with Life under the effect of α-asynchronous perturbation. Basically, we are asking the fundamental question again and again – what is there so special about the Life rules? Moreover, this study displays the gallery of special extended-Life games (identified by Torre and Mártin [13]) under α-asynchronous perturbation. Finally, the study concludes with some random games with different peculiar phase transition dynamics. In this scenario, the next section introduces the Life and α-asynchronous perturbation.

2   Life, α-Asynchronism and Experimental Setup

Traditionally, Life evolves in a regular subset of $\mathbb{Z}^2$ with state set $S = \{0, 1\}$ (on other words, $S = \{dead, alive\}$). In this experimental study, we consider configuration of finite squares with $N \times N$ cells under periodic boundary condition, i.e. $\mathbb{Z}/N\mathbb{Z}$. Life shows dramatic change in dynamics for other (specifically, open) boundary condition [9]. Life follows *Moore* neighbourhood dependency, i.e. self and eight nearest neighbours. Following are the local transition rule of Life.

> Birth rule: A dead cell with exactly three live neighbours evolves to live state. Here, we denote by B3 where B represents birth; and
> Survival rule: A live cell with exactly two or three live neighbours will remain alive. We denote by S23 where S depicts survival.

To sum up, Life rule can be written as B3/S23. In general, Bp/Sq is traditionally used for the naming of these 2-D outer-totalistic games where p and q are the subsets that can contains digits from 0 to 8 to represent the number of live neighbours. Note that, we follow this naming approach to represent Life-like games and extended Life games.

Traditionally, like other synchronous systems Life also assume a global clock that forces the cells to get updated simultaneously. However, the assumption of global clock is not very natural. In this study, for Life and Life-like games we follow α-asynchronous updating scheme [11] where each cell is updated with probability α at each time step, on other words, each cell is left unchanged with probability $1 - \alpha$ at each time step. Note that, $\alpha = 1.0$ depicts the traditional synchronous dynamics. On the other hand, for $\alpha = 0.01$, the system shows almost same dynamics as fully asynchronous perturbation[1].

---

[1] For fully asynchronous updating, one cell is chosen uniformly at random and updated at each time step, and the other cells are left unchanged [28, 31].

Next, to understand the effect of α-asynchronous perturbation, we consider the following qualitative and quantitative experimental setup.

- Firstly, in the qualitative experiment, we evolve Life and Life-like games starting with 50×50 ($N \times N$) random initial configuration of fixed density (no of alive cell). During the evolution of the system, we need to observe the space-time digram of Life and Life-like games. In this experimental approach, we can able to provide a visual comparison between Life and Life-like games under α-asynchronous perturbation.
- Secondly, quantitative experiment follows the well-known approach of [16, 27]. Here, in the quantitative (formal) experiment, we observe the density (say, d_x) of configuration (say, x), i.e. $d_x = \frac{x_{alive}}{|x|}$ where x_{alive} counts the number of alive cell in the configuration x; and $|x|$ is the size of the configuration space (in this study, 50×50). In this quantitative experiment, we evolve the Life or Life-like games for a transient time period (say, $t_{transient}$) starting from an initial configuration with density d_{ini}. Here, in this study, $t_{transient} = 1000$

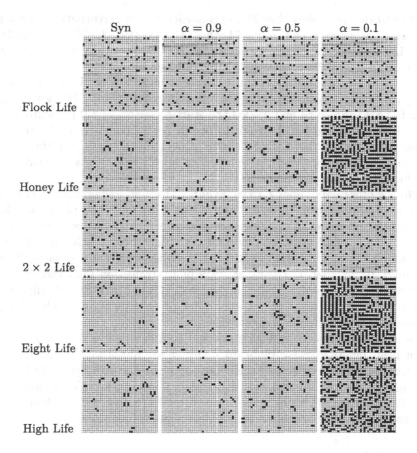

Fig. 2. Space-time diagrams of low density Life-like games after transient time steps.

time steps. Next, we calculate the average density of Life or Life-like games for a sampling period of time (say, $t_{sampling}$). We denote this average density by d_{avg}. In this study, $t_{sampling} = 100$ time steps. To sum up, $d_{avg}(d_{ini}, \alpha)$ depicts the steady-state density under the proposed α-asynchronous perturbation environment. In general, we consider $d_{ini} = 0.5$.

Following this experimental setup, we next explore – (a) low density Life-like games, like, Eight Life, Pedestrian Life, Flock Life [14,18,19,26]; (b) high density Life-like games, like, Dry Life, Drigh Life [14,19,26]; (c) extended-Life games capable to generate natural patterns like, 'labyrinths', 'island with borders', 'solid-liquid mixture' [13]; (d) random games capable to show different kind of phase transitions; under α-asynchronous perturbation.

3   Life-Like Games and Perturbation

3.1   Low-Density Life-Like Games and Perturbation

In this section, we explore the effect of α-asynchronism perturbation on low density Life-like games, specifically – Flock Life (B3/S12), 2×2 Life (B36/S125), High Life (B36/S23), Pedestrian Life (B38/S23), Eight Life (B3/S238), Honey Life (B38/S238), LowDeath Life (B368/S238).

Let us first focus on the qualitative experimental results. Here, Fig. 2 depicts the space-time diagrams after transient time steps for Flock Life, Honey Life, 2×2 Life, Eight Life and High Life starting with $\alpha = 1.0$ (synchronous) to $\alpha = 0.1$. According to Fig. 2, Flock Life and 2×2 Life show a solid resistance against perturbation, i.e. *inactive sparse* phase remains same with the changing value of $\alpha \in [0,1]$. For quantitative evidence, see the steady-state density profile as a function of α perturbation in Fig. 3.

On the other hand, Honey Life, Eight Life, and Pedestrian Life depict a phase transition from *inactive sparse* phase to *labyrinth* phase for changing value of α. For evidence, Fig. 2 depicts the *inactive sparse* phase for $\alpha = 1.0(syn), 0.9, 0.5$ and *labyrinth* phase for vary high[2] perturbation rate ($\alpha = 0.1$) considering Honey Life and Eight Life. Note that, Life also depicts the same dynamics. Recall that, Life also shows a phase transition from *inactive sparse* phase to *labyrinth phase* for changing value of α, see Fig. 1. Specifically, according to [15,16], Life displays a second-order phase transition which belongs to the directed percolation universality class.

In a different dynamics, High Life and LowDeath Life show a phase transition from *inactive sparse* phase to *active dense* phase for increasing rate of perturbation, for evidence, see the space-time diagram of High Life in Fig. 2. Figure 3 depicts the quantitative experimental results for all the low-density Life-like games where High Life, Pedestrian Life, Eight Life, Honey Life, LowDeath Life show the interesting phase transition dynamics.

---

[2] Note that, here, high (resp. low) perturbation rate means low (resp. high) α-value.

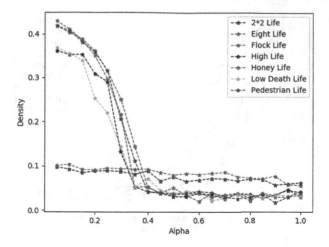

Fig. 3. The plot depicts the density parameter profile as a function of alpha parameter for lower density Life-like games.

3.2   High-Density Life-Like Games and Perturbation

Here, we discuss the effect of the α-asynchronous perturbation on high-density Life-like games, specifically – Dry Life (B37/S23), Drigh Life (B367/S23), B356/S23, B356/S238, B3568/S23, B3568/S238, B3578/S23, B3578/S237, B3578 /S238.

As a qualitative result, Fig. 5 shows the samples of space-time diagrams after transient time steps for Dry Life (B37/S23), Drigh Life (B367/S23), B356/S238 and B3578/S238 starting from $\alpha = 1.0$ (synchronous) to $\alpha = 0.1$. On the other hand, Fig. 4 depicts the quantitative experimental steady-state density profiles as a function of alpha perturbation for all high-density Life-like games. According to Fig. 4, all the high-density Life-like games show the phase transition dynamics with the effect of alpha-perturbation. Therefore, phase transition is a general dynamics for these games with the effect of noise.

Here, only Dry Life (B37/S23) shows (out of high-density Life-like games) a phase transition from *inactive-sparse* phase to *labyrinth* phase for changing value of α perturbation, for evidence see Fig. 5. Recall that, Life also shows the similar phase change towards *labyrinth* phase with the effect of α perturbation [15,16]. However, the rest high-density Life-like games depict a phase transition from *inactive sparse* phase to active dense phase for changing value of α perturbation, see Fig. 5 for evidence. Therefore, only Dry Life follows the dynamics of Life. In this context, note that, Life also shows phase transition from *inactive sparse* phase to *active dense* phase with the effect of probabilistic loos of information perturbation [27].

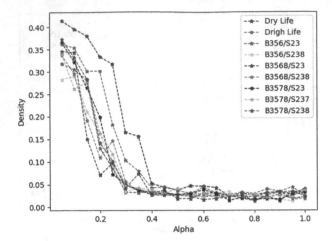

Fig. 4. The plot depicts the density parameter profile as a function of alpha parameter for higher density Life-like games.

3.3   Extended Life Games and Perturbation

In a early work, Torre and Mártin [13] have explored all 1296 *extended* [3] Life games and have identified (some of) their capability of generating natural patterns like 'labyrinths', 'island with borders', 'ferromagnetic domains', 'solid-liquid mixture'. In this section, we specifically explore those special (capable to generate natural patterns) extended Life games under α-asynchronous environment.

According to Torre and Mártin [13], games like B3/S1234, B345678/S56, B12/S1234 show *labyrinths* structure under traditional synchronous environment. These rules show solid resistance against the perturbation. For evidence, Fig. 6 depicts the dynamics of B12/S1234 under α-asynchronous environment. On the other hand, B123/S1234 shows *labyrinths* pattern for high perturbation rate, $\alpha = 0.1, 0.5$, see Fig. 6. However, the game B123/S1234 shows dynamics like *ferromagnetic* domains under synchronous environment. Note that, games B4/S1234 and B1234567/S56 also show the same dynamics (like B123/S1234). Under synchronous environment, game B234567/S5678 shows *phases in combat* where different patterns are generated in few region of the lattice starting with small initial density, hereafter, those pattern grows until the complete coverage of the lattice. For α-asynchronous environment, one can also observe the *phases in combat*, however, finally, the game converges and shows pattern like 'large building with some light-on windows during late night', see Fig. 6 for evidence. Here,

---

[3] Following the naming Bp/Sq, we can write $p = [p_1, p_2]$ and $q = [q_1, q_2]$ where the parameters take the number of live neighbours within the interval p_1 and p_2; resp. q_1 and q_2. Considering parameters p_1, p_2, q_1, q_2, one can take integer values from 1 to 8, and there are 1296 games. Here, we call those games as extended Life games [13].

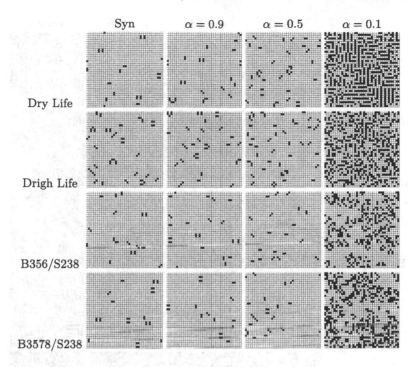

Fig. 5. Space-time diagrams of high density Life-like games after transient time steps.

B345678/S5678 and B34567/S567 also depicts the same dynamics. According to [13], game B45678/S5678 shows islands with active borders starting with initial density $d_{ini} = 0.5$ under synchronous environment. However, the size of islands decrease with the decreasing value α. Moreover, B45678/S5678 also depicts 'large building with some light-on windows during late night' pattern for high rate of perturbation, for $\alpha = 0.1$ see Fig. 6. Here also, game B45/S456 shows the similar dynamics. Similarly, under the synchronous environment, B45678/S4567 shows pattern like 'islands with overlapping lake inside' starting with initial density $\alpha = 0.3$. This game shows resistance against small perturbation, for $\alpha = 0.9$ see Fig. 6. However, for high perturbation rate (i.e. $\alpha = 0.1$), game B45678/S4567 shows the similar 'large building with some light-on windows during late night' dynamics. To sum up, Fig. 6 displays the beautiful gallery of these rules with the effect of perturbation. Note that, this part of study on extended Life games only focuses on the qualitative experiment.

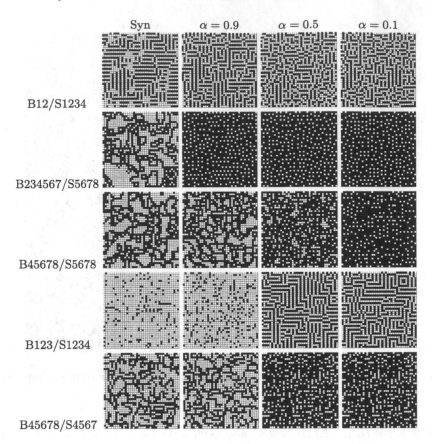

Fig. 6. Space-time diagrams of extended Life games after transient time steps.

4   Discussion

To sum up, this first experiment on Life-like cellular automata with the effect of α-asynchronous perturbation shows following peculiar behaviour –

- Low density Flock Life and 2 × 2 Life show a solid resistance against perturbation.
- Low density games like Honey Life, Eight Life and Pedestrian Life show similarity in dynamics with Life. With the effect of perturbation, these games show phase transition towards labyrinth patterns.
- Low density games like High Life and LowDeath Life also depict phase transition dynamics, however, for high rate of perturbation, these games show active-dense pattern.
- All high density Life-like games show phase transition dynamics with the effect of perturbation. However, only Dry Life follows similar dynamics with Life, i.e. labyrinth patterns.

– Extended `Life` games with labyrinth patterns also show solid resistance against perturbation. However, most of natural patterns of extended `Life` games show stability with the effect of α-asynchronous perturbation.

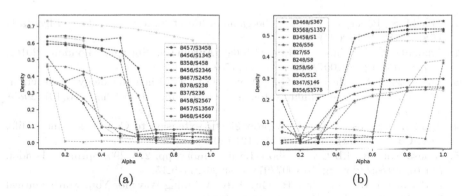

(a)   (b)

Fig. 7. The plot depicts the density parameter profile as a function of alpha parameter for random games.

According to all these evidence, we can say that phase transition is a general trend for `Life`-like games under α-asynchronous environment. For more evidence, in a first experiment, we explore 1000 random outer-totalistic games and 137 (out of 1000) games show different kind of phase transition dynamics with the effect of perturbation. Figure 7 displays some of the quantitative experimental results associated with phase transition dynamics. Figure 7(a) shows games which show almost zero density (inactive-sparse phase) for traditional synchronous environment and depicts active dense phase ($d_{avg} > 0.2$) after a certain perturbation rate. Here, game B457/S13567 shows very early (low perturbation rate) phase transition for critical value $\alpha_c \sim 0.9$, on the other hand, game B458/S2567 depicts very late ($\alpha_c \sim 0.2$) phase transition. Some games like B456/S1345, B468/S4568 show phase transition for $\alpha_c \sim 0.5$, see Fig. 7(a) for all these various behaviour.

In other interesting dynamics, game like B3458/S1, B258/S6, B248/S8, B26/ S56 show a phase transition from active-dense phase to extinct phase (zero-density), see Fig. 7(b). Here also, B3458/S1 shows early phase transition ($\alpha_c \sim 0.9$) and games like B258/S6, B248/S8, B26/S56 depict late phase transition ($\alpha_c \sim 0.2$). In this context, note that, during the evolution of stochastic `Life` rules, Monetti and Albano [23,24] have also observed first-order irreversible phase transition from steady state (non-zero density) to extinct phase (zero density). Lastly, games like B356/S3578, B3468/S367 show two phase transitions – firstly, form active-dense to almost zero density; secondly, from almost zero density to again active-dense for very high rate of perturbation, see Fig. 7(b) for all these peculiar dynamics. Therefore, phase transition dynamics with the effect of perturbation is also visible for many Bp/Sq 2-D outer-totalistic rules (along with

`Life`). However, the connection between a random game (the set p and q) and phase transition is still open to us.

References

1. Adachi, S., Peper, F., Lee, J.: The game of life at finite temperature. Physica D **198**(3), 182–196 (2004)
2. Adamatzky, A.: Game of Life Cellular Automata. Springer, London (2010). https://doi.org/10.1007/978-1-84996-217-9
3. Adamatzky, A., Martínez, G.J., Mora, J.C.S.T.: Phenomenology of reaction-diffusion binary-state cellular automata. Int. J. Bifurcation Chaos **16**(10), 2985–3005 (2006)
4. Aguilera-Venegas, G., et al.: A probabilistic extension to Conway's game of life. Adv. Comput. Math. **45**(4), 2111–2121 (2019)
5. Alonso-Sanz, R.: LIFE with Short-Term Memory, pp. 275–290. Springer, London (2010). https://doi.org/10.1007/978-1-84996-217-9_15
6. Berlekamp, E.R., Conway, J.H., Guy, R.K.: Winning Ways for Your Mathematical Plays, vol. 2. Academic Press, London (1984)
7. Bersini, H., Detours, V.: Asynchrony induces stability in cellular automata based models. In: Artificial Life IV: Proceedings of the Fourth International Workshop on the Synthesis and Simulation of Living Systems, pp. 382–387. The MIT Press, Cambridge (1994)
8. Bleh, D., Calarco, T., Montangero, S.: Quantum game of life. Europhys. Lett. **97**(2), 20012 (2012)
9. Blok, H.J., Bergersen, B.: Effect of boundary conditions on scaling in the "game of life." Phys. Rev. E **55**, 6249–6252 (1997)
10. Blok, H.J., Bergersen, B.: Synchronous versus asynchronous updating in the "game of life." Phys. Rev. E **59**, 3876–3879 (1999)
11. Bouré, O., Fatès, N.A., Chevrier, V.: Probing robustness of cellular automata through variations of asynchronous updating. Nat. Comput. **11**(4), 553–564 (2012)
12. Das, S., Roy, S., Bhattacharjee, K.: The Mathematical Artist: A Tribute To John Horton Conway. Emergence, Complexity and Computation, Springer, Cham (2022). https://doi.org/10.1007/978-3-031-03986-7
13. de la Torre, A.C., Mártin, H.O.: A survey of cellular automata like the "game of life." Phys. A **240**(3), 560–570 (1997)
14. Eppstein, D.: Growth and Decay in Life-Like Cellular Automata, pp. 71–97. Springer, London (2010). https://doi.org/10.1007/978-1-84996-217-9_6
15. Fatès, N.: Does Life Resist Asynchrony?, pp. 257–274. Springer, London (2010). https://doi.org/10.1007/978-1-84996-217-9_14
16. Fatès, N., Morvan, M.: Perturbing the topology of the game of life increases its robustness to asynchrony. In: Sloot, P.M.A., Chopard, B., Hoekstra, A.G. (eds.) ACRI 2004. LNCS, vol. 3305, pp. 111–120. Springer, Heidelberg (2004). https://doi.org/10.1007/978-3-540-30479-1_12
17. Gardner, M.: Mathematical games: the fantastic combinations of john Conway's new solitaire game "life." Sci. Am. **223**(4), 120–123 (1970)
18. Johnston, N.: The B36/S125 "2x2" Life-Like Cellular Automaton, pp. 99–114. Springer, London (2010). https://doi.org/10.1007/978-1-84996-217-9_7
19. LifeWiki. List of life-like cellular automata (2021)

20. Magnier, M., Lattaud, C., Heudin, J.-C.: Complexity classes in the two-dimensional life cellular automata subspace. Complex Syst. **11**, 419–436 (1997)

21. Martínez, G.J., Adamatzky, A., Seck-Tuoh-Mora, J.C.: Some notes about the game of life cellular automaton. In: Das, S., Roy, S., Bhattacharjee, K. (eds.) The Mathematical Artist. Emergence, Complexity and Computation, vol. 45, pp. 93–104. Springer, Cham (2022). https://doi.org/10.1007/978-3-031-03986-7_4

22. Jaroslaw Adam Miszczak: Rule switching mechanisms in the game of life with synchronous and asynchronous updating policy. Phys. Scr. **98**(11), 115210 (2023)

23. Monetti, R.A.: First-order irreversible phase transitions in a nonequilibrium system: mean-field analysis and simulation results. Phys. Rev. E **65**, 016103 (2001)

24. Monetti, R.A., Albano, E.V.: Critical edge between frozen extinction and chaotic life. Phys. Rev. E **52**, 5825–5831 (1995)

25. Ney, P.-M.., Notarnicola, S., Montangero, S., Morigi, G.: Entanglement in the quantum game of life. Phys. Rev. A **105**, 012416 (2022)

26. Peña, E., Sayama, H.: Life worth mentioning: complexity in life-like cellular automata. Artif. Life **27**(2), 105–112 (2021)

27. Roy, S.: A study on delay-sensitive cellular automata. Phys. A **515**, 600–616 (2019)

28. Roy, S.: Asynchronous cellular automata that hide some of the configurations during evolution. Int. J. Mod. Phys. C **32**, 2150054 (2021)

29. Roy, S., Paul, S., Das, S.: Temporally stochastic cellular automata: classes and dynamics. Int. J. Bifurcation Chaos **32**(12), 2230029 (2022)

30. Schulman L.S., Seiden, P.E.: Statistical mechanics of a dynamical system based on Conway's game of life. J. Stat. Phys. **19**(3), 293–314 (1978)

31. Sethi, B., Roy, S., Das, S.: Asynchronous cellular automata and pattern classification. Complexity **21**(S1), 370–386 (2016)

32. Turing, A.: The chemical basis of morphogenesis. Bull. Math. Biol. **52**, 153–197 (1952)

On Elementary Second Order Cellular Automata

Enrico Formenti[1]($\boxtimes$)(iD) and Supreeti Kamilya[2]($\boxtimes$)(iD)

[1] Université Côte d'Azur, CNRS, I3S, Nice, France
enrico.formenti@univ-cotedazur.fr
[2] Department of Computer Science and Engineering, Birla Institute of Technology,
Mesra, Ranchi, India
kamilyasupreeti779@gmail.com

Abstract. The paper performs a first exploration of elementary higher order cellular automata. As a first result we give the index of the equivalence relation induced by the usual set of transformations on the local rule (adapted to the new context) when the order is 2. This reveals that the class of second order CA is to large for the purposes of our paper. Hence, we propose to add a further constraint on the structure of the local rule. We call this new class RESOCA . Finally, we provide several results about the dynamics of a large set of RESOCA .

1 Introduction

Classically, a **cellular automaton** (CA) is a discrete dynamical model comprising a regular lattice of simple computing units called **cells**. Each cell has a **state** chosen from a finite set S called the **set of states** or the **alphabet**. These cells simultaneously update their states in discrete time steps based on a uniform local rule and the present states of a set of neighboring cells called **the neighborhood**. The neighborhood is also uniform (*i.e.* it has the same structure for all cells). The simplicity and regularity inherent in the architecture of CA can lead to the emergence of remarkably complex and irregular patterns during their evolution. This characteristic motivates scientists to extensively explore them across various domains, including mathematics, physics, and computer science. In early 1950s, upon receiving input from Stanislaw Ulam, John Von Neumann employed a regular lattice as a mathematical representation for self-reproduction. Since then, they found application in almost any scientific domain.

We remark that the CA model is memoryless in the sense that the new state of cells is computed only on the basis of the current state of the system. Indeed, there are situations in which this is a drawback. Hence, researchers started turning their attention to CA which can hold some memory about a (finite) number of past states. Building on seminal papers of Toffoli [15], Le Bruyn, Van Den Bergh and Kari provided the first characterizations of injectivity, surjectivity and bijectivity [3,10]. More recently, an complete and decidable characterization

M. Dalui et al. (Eds.): ASCAT 2024, CCIS 2021, pp. 204–218, 2024.
https://doi.org/10.1007/978-3-031-56943-2_16

of the dynamical behavior of linear higher order CA has been provided [7,8,11]. In this paper we follow this trend investigating a particular class of second order CA induced by classical elementary CA.

The paper is organized as follow. The next section introduces preliminaries about elementary high order CA (EHOCA) as well as the standard elementary CA, and the main notions about the dynamics of EHOCA that are needed in the paper. Section 3 presents results about transformations on local rules of EHOCA. In particular, it shows that even if one takes the quotient of the rule space with respect to the given set of transformations, the number of non-equivalent classes is too huge to be tackled in the present paper. Therefore, we present RESOCA , an interesting sub-class of EHOCA in Sect. 4 and provide several results about their dynamics. In the last section we draw our conclusions and present some directions for future research.

Due to lack of space some proofs have been moved into the appendix, they will appear in the long version of the paper.

2   Preliminaries

2.1   Elementary Higher Order CA

Formally, an **elementary higher order CA** (EHOCA) is a tuple $\langle S, m, N, \delta \rangle$ where $S = \{0,1\}$ is the **set of states**; m is the **order** (sometimes also called the **memory**); $N \colon \mathbb{Z} \to \left(\{0,1\}^3 \right)^m$ is the **neighborhood** function which associates each cell with the content of the nearest neighboring cells *w.r.t.* the current time step and to the past one; finally, $\delta \colon \left(\{0,1\}^3 \right)^m \to \{0,1\}$ is the **local rule** which updates the state of the current cell according to the content of the neighborhood and of the memory. Let us denote Δ the space of local rules of EHOCA. A **configuration** is a snapshot of the state of all the finite automata *i.e.* a function from $\mathbb{Z}$ to $\{0,1\}^m$ (recall that m is the memory size). Denote $\left(\{0,1\}^{\mathbb{Z}} \right)^m$ the set of all configurations. For $m = 1$, we will denote $\underline{1}$ the configuration which has a 1 at cell of index 0, and 0 elsewhere; similarly, $\underline{0}$ is the configuration in which all cells are in state 0. If $G_\delta \left((\underline{0})^m \right) = (\underline{0})^m$, the state $(0)^m$ is called **quiescent**. Configurations for which only a finite number of cells are not in quiescent state are called **finite configurations**.

A local function δ induces a **global function** $G_\delta \colon \left(\{0,1\}^{\mathbb{Z}} \right)^m \to \left(\{0,1\}^{\mathbb{Z}} \right)^m$ describing how the current configuration evolves in the next time step. In case of an elementary second order CA, the value of m is considered as 2. More precisely, for elementary second order CA, G_δ associates the vector $\boldsymbol{e} = (e^1, e^2) \in \left(\{0,1\}^{\mathbb{Z}} \right)^2$ with the vector $G_\delta(\boldsymbol{e}) \in \left(\{0,1\}^{\mathbb{Z}} \right)^2$ where $G_\delta(\boldsymbol{e})_i^1 = e_i^2$ and

$$G_\delta(\boldsymbol{e})_i^2 = \delta \left(\begin{bmatrix} (e_{i-1}^1, e_i^1, e_{i+1}^1) \\ (e_{i-1}^2, e_i^2, e_{i+1}^2) \end{bmatrix} \right)$$

for all $i \in \mathbb{Z}$. In other words, the new vector at position i is obtained by taking the old value of component 2 and putting it in component one, and computing the updated value of the new component 2 using the local rule δ.

In order to estimate the distance between two configurations, one can introduce the following metric:

$$\forall x, y \in \left(\{0,1\}^{\mathbb{Z}}\right)^m \quad d(x,y) = 2^{-k} \tag{1}$$

where $k = \min \{i \in \mathbb{Z} \mid x_{|i|} \neq y_{|i|}\}$. When the set of configuration is equipped with the topology induced by the metric in (1), then the global function of EHOCA is a continuous function and the structure $\left\langle \left(\{0,1\}^{\mathbb{Z}}\right)^m, G_\delta \right\rangle$ is a dynamical system and the topology induced by d coincides with the classical Cantor topology. Specifically, the **Cantor topology** is derived by utilizing the product topology on the set of states of the automata, which is equipped with the discrete topology. A Cantor space is defined as a metric space that is compact, signifying that any sequence within it possesses a convergent subsequence. It is also totally disconnected, with distinct points being separated by disjoint clopen sets (*i.e.* sets that are both closed and open). Additionally, a Cantor space is perfect, meaning that no point within it is isolated. Notably, any two Cantor spaces are homeomorphic.

In the context of discrete dynamical theory, researchers are interested in the study of the long-term behavior of the systems. We recall here the definitions concerning some of the behaviors that we are directly concerned with in this paper. For convenience sake, the notions are specialized to the context of EHOCA.

A configuration $c \in (\{0,1\}^{\mathbb{Z}})^m$ of an EHOCA $\langle \{0,1\}, m, \{-1,0,+1\}, \delta \rangle$ is **periodic** with **period** $n > 0$, if $G_\delta^n(c) = c$. It is **eventually periodic**, if there exists some **preperiod** $m > 0$ and some period n such that $G_\delta^{m+kn}(c)$ is periodic for all $k \in \mathbb{N}$. G_δ is periodic if all configurations are periodic. A configuration $c \in (\{0,1\}^{\mathbb{Z}})^m$ is an **equicontinuity point** for G_δ if

$$\forall \epsilon > 0, \exists \delta > 0, \forall c' \in (\{0,1\}^{\mathbb{Z}})^m, d(c',c) < \delta \implies \forall t \geq 0 : d(G_\delta^t(c'), G_\delta^t(c)) < \epsilon$$

We represent the set of all equicontinuity points of G_δ by $eq(G_\delta)$. A EHOCA G_δ is **equicontinuous** if $eq(G_\delta) = (\{0,1\}^{\mathbb{Z}})^m$, *i.e.*, all configurations are equicontinuity points. **Sensitivity to initial conditions** is a somewhat logically opposite property than equicontinuity and it is often seen as an element of unpredicatibility of the system. An HOCA is sensitive to initial conditions iff $\exists \epsilon > 0 \forall x \in (\{0,1\}^{\mathbb{Z}})^m \forall \xi > 0 \exists y \in B_\xi(c)$ such that $\forall t \geq 0 : d(G_\delta^t(x), G_\delta^t(x)) < \epsilon$, where $B_\xi(c)$ is the open ball of radius ξ centered in c. Expansivity is a stronger form of sensitivity to initial conditions. A EHOCA is **(positively) expansive**

$$\exists \epsilon > 0 \forall x, y \in (\{0,1\}^{\mathbb{Z}})^m \exists t \in \mathbb{N} \; d(G_\delta^t(x), G_\delta^t(y)) \geq \epsilon.$$

A EHOCA **strongly transitive** if for any open set $U \neq \emptyset$, $\bigcup_{t \in \mathbb{N}} G_\delta^t(U) = (\{0,1\}^{\mathbb{Z}})^m$. Strong transitivity if often seen as strong indecomposability of the system *i.e.* the system cannot be decomposed into non-trivial disjoint subsystems. A EHOCA has the dense periodic orbits **DPO property** if its set of periodic points is dense in the space of configurations.

An EHOCA is **injective** if its global transition function G_δ is one-to-one. If the function is onto, the EHOCA is **surjective**. An EHOCA is **bijective** if G_δ is both one-one and onto. We recall that for any EHOCA, there exists a standard CA which is topologically conjugated to it [8]. Hence, we call a EHOCA **reversible** if and only if its conjugated CA is reversible. By a result in [9], the class of reversible EHOCA coincides with the class of bijective EHOCA.

2.2 Elementary Cellular Automata (ECA)

A EHOCA is an **elementary CA** (or ECA) if memory $m = 1$. Hence, the local rule δ can be conveniently represented in tabular form (see for example Table 1). Interpreting the output row of the tabular representation of a local rule as a decimal number written in binary gives the well-known **Wolfram's number**. Indeed, in the sequel of this paper we will refer to a specific ECA by its Wolfram's number n and refer to its local rule by δ_n.

There are a total of 256 distinct ECA rules although from the discrete dynamical systems point of view some of the are topologically conjugated via local transformations similar to those given in the previous section. Table 2 shows the minimal non-equivalent ECA rules.

For many years, scientists worldwide have involved themselves in investigating the classical CA (and sometimes just ECA) and categorized them based on the different behaviours of the dynamics [4,6,12,14]. According to the degree of equicontinuity and expansivity of CAs, Kůrka subdivided CAs into four classes in [12]:

K1 Equicontinuous CAs.
K2 Almost equicontinuous but not equicontinuous CAs.
K3 Sensitive but not positively expansive CAs.
K4 Positively expansive CAs.

In [2], Cattaneo *et al.* provided an interesting classification based on pattern growth for ECA which has somewhat inspired the study in the Sect. 4.

Table 1. The table representation of the local rule for ECA 22.

	111	110	101	100	011	010	001	000
f	0	0	0	1	0	1	1	0

3   Transformations on EHOCA Local Rules

By the definition of the local rule it is clear that there are 2^{2^6} distinct EHOCA. Hence, studying all of them individually is a challenging task. However, many of the dynamics of EHOCA are topologically conjugated *i.e.* they are equivalent from the point of view of topological dynamics.

Table 2. ECA local rules up to local transformations in $\mathcal{T}$.

0, 1, 2, 3, 4, 5, 6, 7, 8, 9, 10, 11, 12, 13, 14, 15, 18, 19, 22, 23, 24, 25, 26, 27, 28, 29, 30, 32, 33, 34, 35, 36, 37, 38, 40, 41, 42, 43, 44, 45, 46, 50, 51, 54, 56, 57, 58, 60, 62, 72, 73, 74, 76, 77, 78, 90, 94, 104, 105, 106, 108, 110, 122, 126, 128, 130, 132, 134, 136, 138, 140, 142, 146, 150, 152, 154, 156, 160, 162, 164, 168, 170, 172, 178, 184, 200, 204, 232

Two dynamical systems $\langle X, f \rangle$ and $\langle Y, g \rangle$ are **topologically conjugated** if and only if there exists an homeomorphism (*i.e.* a continuous bijective map whose inverse is also continuous) ϕ such that $\phi \circ f = g \circ \phi$ *i.e.* the following diagram commutes

$$
\begin{array}{ccc}
X & \xrightarrow{\ f\ } & X \\
\phi \downarrow & & \downarrow \phi \\
Y & \xrightarrow[\ g\ ]{} & Y
\end{array}
$$

As already observed in [8], any EHOCA $\langle \{0,1\}, 2, N, \delta \rangle$ is topologically conjugate to a classical CA with set of states $S = \{0,1\}^2$, radius $r = 1$ and a suitable local rule f. A **transformation on local rules** is a mapping from the space of EHOCA local rules to itself. Inspired by [5], in the context of EHOCA, one can consider the following transformations on local rules:

$$
\tau_r(\delta)\left(\begin{bmatrix} a, b, c \\ d, e, f \end{bmatrix}\right) = \delta\left(\begin{bmatrix} c, b, a \\ f, e, d \end{bmatrix}\right)
$$

$$
\tau_n(\delta)\left(\begin{bmatrix} a, b, c \\ d, e, f \end{bmatrix}\right) = 1 - \delta\left(\begin{bmatrix} 1-a, 1-b, 1-c \\ 1-d, 1-e, 1-f \end{bmatrix}\right)
$$

$$
\tau_{rn}(\delta)\left(\begin{bmatrix} a, b, c \\ d, e, f \end{bmatrix}\right) = \tau_r \circ \tau_n = 1 - \delta\left(\begin{bmatrix} 1-c, 1-b, 1-a \\ 1-f, 1-e, 1-d \end{bmatrix}\right)
$$

It is clear from the definitions that $\tau_{rn} = \tau_{nr} = \tau_n \circ \tau_r$. Consider the following set of local rule transformations $\mathcal{T} = \{\tau_i, \tau_r, \tau_n, \tau_{rn}\}$ and define the following equivalence $\Re$ over Δ:

$$
\forall \delta_1, \delta_2 \in \Delta \quad \delta_1 \Re \delta_2 \iff \exists \tau \in \mathcal{T} \text{ s.t. } \tau(\delta_1) = \delta_2
$$

A local rule $\delta \in \Delta$ is a **fixed point** for a transformation τ iff $\tau(\delta) = \delta$. We denote Fix(τ) the set of fixed points of the transformation $\tau \in \mathcal{T}$. A transformations τ over a set T is **self-inverse** if $\tau \circ \tau = id$, where id is the identity mapping over T. It not difficult to see that all transformations in $\mathcal{T}$ are self-inverse. **Transformation graphs** are a useful way to represent transformations on local rules. All transformation graphs have the same vertex set, namely, 3×3 matrices over $\{0, 1\}$ while the edge set depends on the local transformation. Here are the four that we are concerned with in this paper:

$$E_{\tau_i} = \left\{ \left(\begin{bmatrix} a, b, c \\ d, e, f \end{bmatrix}, \begin{bmatrix} a, b, c \\ d, e, f \end{bmatrix} \right) \mid a, b, c, d, e, f \in \{0, 1\} \right\}$$

$$E_{\tau_r} = \left\{ \left(\begin{bmatrix} a, b, c \\ d, e, f \end{bmatrix}, \begin{bmatrix} c, b, a \\ f, e, d \end{bmatrix} \right) \mid a, b, c, d, e, f \in \{0, 1\} \right\}$$

$$E_{\tau_n} = \left\{ \left(\begin{bmatrix} a, b, c \\ d, e, f \end{bmatrix}, \begin{bmatrix} 1-a, 1-b, 1-c \\ 1-d, 1-e, 1-f \end{bmatrix} \right) \mid a, b, c, d, e, f \in \{0, 1\} \right\}$$

$$E_{\tau_{rn}} = \left\{ \left(\begin{bmatrix} a, b, c \\ d, e, f \end{bmatrix}, \begin{bmatrix} 1-c, 1-b, 1-a \\ 1-f, 1-e, 1-d \end{bmatrix} \right) \mid a, b, c, d, e, f \in \{0, 1\} \right\}$$

Edges (U, V) in a transformation graph G_{τ_i} or G_{τ_r} of a local rule δ are labelled $\delta(V)$ while for G_{τ_n} or $G_{\tau_{rn}}$ are labelled $1 - \delta(V)$. It is clear from the definitions that the transformation graphs that we just defined are made of cycles of size 2 and loops. It is also clear that a local rule is a fixed point of a transformation τ if for any cycle the labels of its edges are identical (but different cycles might have different labels of course). This simple remark will reveal to be very important when counting the number of fixed points of local transformations.

Lemma 1. $|\text{Fix}(\tau_r)| = 2^{24}$.

Proof. As already remarked earlier, in order to have a fixed point, it is clear that the labels of the edges in the same cycle must be identical. Here, we shall consider the graph of the transformation τ_r. First, consider the components in which the 'upper' part of e is $(0, 0, 0)$ (as shown in Fig. 1). There can be a total of 6 such graphs. From each component, we should have two possible outcomes for the labels, 0 or 1. Therefore, the number of fixed points is 2^6 when the 'upper' part is $(0, 0, 0)$. Remark that the graph is the same if the upper part is chosen in $\{(0, 1, 0), (1, 0, 1), (1, 1, 1)\}$ except for the labels of the edges, of course. Therefore, for those four subgraphs of τ_r shown in the figure, we obtain 4×2^6 number of fixed points.

Now, let us consider the components of the transformation τ_r when the 'upper' part is $(0, a, 1)$ or $(1, a, 0)$, for $a, b, c \in \{0, 1\}$ and $b = 1 - c$ (as shown in Fig. 2). The 'upper' part of the first graph of Fig. 2 can have four combinations – $(0, a, 0)$, $(0, a, 1)$, $(1, a, 0)$ and $(1, a, 1)$. Similarly, there are four combinations for each of the graphs shown in Fig. 2. So, the total number of possible graphs is $4 \times 4 = 16$. Two outcomes (0 or 1) are possible for the labels of each of them. Therefore, the number of fixed points is 2^{16}. Now, we understand that, for 2^{16} possible components, we can have 4×2^6 fixed points in each. Hence, the total number of fixed points is, $4 \times 2^6 \times 2^{16} = 2^{24}$.

Lemma 2. $|\text{Fix}(\tau_n)| = 2^{32}$.

Proof. It is not difficult to see that the graph of the transformation τ_n with $(a, b, c) \in \{(0, 0, 0), (0, 0, 1), (0, 1, 1), (1, 0, 0)\}$ and $d, e, f \in \{0, 1\}$ is made of cycles as shown in Fig. 3. For each combination of (a, b, c), there are 8 possible combinations in the 'lower' part. So, the total number of such cycles of size 2 is $4 \times 8 = 32$ and each one provides two possibilities for the edge labels. This gives 2^{32} fixed points.

Lemma 3. $|\text{Fix}(\tau_{rn})| = 2^{32}$.

Proof. The proof is very similar to those of previous lemmas. In this case the transformation graph of τ_{rn} with $(a, b, c) \in \{(0, 0, 0), (0, 0, 1), (0, 1, 0), (1, 0, 0)\}$ and $d, e, f \in \{0, 1\}$ is made of cycles of size 2 (as shown in Fig. 4). Since there are 32 such cycles, there are 2^{32} distinct fixed points.

Lemma 4. $\mathcal{T}$ *is an abelian group w.r.t. standard composition of functions* $\circ$.

Proof. It is clear that τ_i, τ_r and τ_n are self-inverse. Indeed, we have that also τ_{rn} is self-inverse:

$$\tau_{rn} \circ \tau_{nr} = (\tau_r \circ \tau_n) \circ (\tau_n \circ \tau_r) = \tau_r \circ (\tau_n \circ \tau_n) \circ \tau_r = \tau_r \circ \tau_r = \tau_i$$

by using the associativity of $\circ$ and the fact that τ_r and τ_n are self-inverse. Let us prove that $\tau_{rn} = \tau_r \circ \tau_n = \tau_{nr}$. Since τ_{rn} is self-inverse we have

$$(\tau_r \circ \tau_n) \circ (\tau_r \circ \tau_n) = \tau_i$$

by pre-multiplying the previous equation by τ_r and using associativity we have

$$\tau_r^2 \circ (\tau_n \circ \tau_r) \circ \tau_n = \tau_r$$

Now, by post-multiplying by τ_n and using the fact that τ_r is self-inverse

$$(\tau_n \circ \tau_r) \circ \tau_n^2 = \tau_r \circ \tau_n$$

and finally using the fact that τ_n is self-inverse, from the previous equation we obtain $\tau_n \circ \tau_r = \tau_r \circ \tau_n$ This last equation allows to conclude that $\mathcal{T}$ is closed w.r.t. $\circ$. The proof that $(\mathcal{T}, \circ)$ is abelian follows directly from $\tau_{rn} = \tau_{nr}$ or from the self-inverse property of the elements of $\mathcal{T}$.

As an immediate consequence of Lemma 2–4 and Burnside's lemma we have the following.

Proposition 1. $|\Delta/\Re| = \frac{|Fix(\tau_i)| + |Fix(\tau_r)| + |Fix(\tau_n)| + |Fix(\tau_{rn})|}{|\Re|} = \frac{2^{64} + 2^{24} + 2 \cdot 2^{32}}{4} = 2^{62} + 2^{31} + 2^{32}$, *where* $Fix(\tau)$ *is the set of fixed points of* τ.

We conclude that the set of EHOCA is too large to be explored class by class. The next section introduces further restrictions which make a first exploration possible.

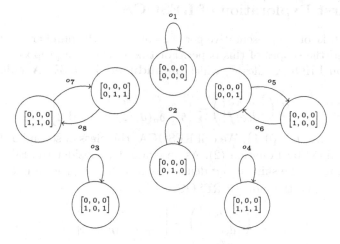

Fig. 1. The graph of the transformation τ_r when the 'upper' part is $0,0,0$. See Lemma 1.

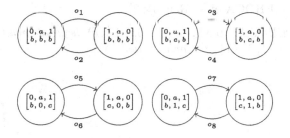

Fig. 2. The graph of the transformation τ_r when the 'upper' part is $(0, a, 1)$ or $(1, a, 0)$ for $a, b, c \in \{0,1\}$ and $b = 1 - c$. See Lemma 1.

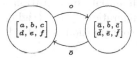

Fig. 3. Graph of the transformation τ_n with $(a, b, c) \in \{(0,0,0), (0,0,1), (0,1,1), (1,0,0)\}$ and $d, e, f \in \{0,1\}$. See Lemma 2.

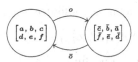

Fig. 4. Graph of the transformation τ_{rn} with $(a, b, c) \in \{(0,0,0), (0,0,1), (0,1,0), (1,0,0)\}$ and $d, e, f \in \{0,1\}$. See Lemma 3.

4   A First Exploration of RESOCA

Even if we take one representative per class of $|\Delta/_\Re|$ the number is too huge and goes beyond the scopes of this paper. For this reason, we propose to focus our attention on EHOCA rules δ'_n which are **induced by a ECA rule δ_n**, more precisely:

$$\delta'_n \left(\begin{bmatrix} a,b,c \\ d,e,f \end{bmatrix} \right) = \begin{bmatrix} e \\ b + \delta_n(d,e,f) \mod 2 \end{bmatrix} \tag{2}$$

for all $a,b,c,d,e,f \in \{0,1\}$. We call RESOCA the class of second order elementary CA defined by Equation (2). As an example, consider the ECA local rule corresponding to the shift map σ defined as $G_\sigma(c)_i = c_{i+1}$ for any configuration $c \in \{0,1\}^{\mathbb{Z}}$. Then, the induced RESOCA σ' is

$$\sigma' \left(\begin{bmatrix} a,b,c \\ d,e,f \end{bmatrix} \right) = \begin{bmatrix} e \\ b + f \mod 2 \end{bmatrix}$$

In Fig. 5, an example of dynamical behavior of the shift map σ is given compared with σ'. The following lemma will allow to adapt easily many CA definitions and properties to the case of EHOCA. The following lemma is a simple adaptation to the context of EHOCA of [8, Remark 2].

Lemma 5. *A EHOCA $\langle \{0,1\}, m, \{-1,0,+1\}, \delta \rangle$ is topologically conjugated to a CA $\langle \{0,1\}^m, \{-1,0,+1\}, \delta' \rangle$ for a suitable δ'.*

Fig. 5. The shift map σ with initial configuration $\underline{1}$ (left) and its corresponding RESOCA σ' with initial configuration $[\underline{1}, \underline{1}]^T$ (right). Time grows downward. Only the second component of each configuration is plotted.

The following proposition provides a first important characterization of RESOCA class that will be widely exploited in the sequel.

Proposition 2. *RESOCA are bijective.*

Proof. As per [13,16], if we define a configuration at time $t+1$ as

$$c_i^{t+1} = \delta(c_{i-1}^t, c_i^t, c_{i+1}^t) + c_i^{t-1} \mod 2,$$

for a given ECA local rule δ, one finds that the above equation has always a solution even if G_δ is not invertible. This makes the induced RESOCA δ' reversible. So, δ' is a bijective CA as any reversible CA is bijective [9] and any RESOCA is topologically conjugated with a particular CA (see Lemma 5).

In order to prove the next result we need the following lemma.

Lemma 6. *Injective CAs have DPO but cannot be expansive or strongly transitive.*

Lemma 7. *Equicontinuous injective CA are periodic.*

As an immediate consequence of Lemma 5, Lemma 6 and Proposition 2 we have the following.

Proposition 3. *RESOCA have DPO but are neither expansive nor strongly transitive. Hence, no RESOCA is in K4.*

Among the 88 ECAs, the additive ones, rule number $0, 60, 90, 150, 170, 204$, can be fully analyzed with the methods reported in [7,8] and hence they are not reconsidered here.

Proposition 4. *The RESOCA induced by ECA rules* $2, 6, 10, 14, 18, 22, 24, 26,$ $28, 30, 34, 38, 42, 46, 50, 54, 56, 58, 62, 74, 78, 94, 106, 110, 122, 126, 130, 134, 138,$ $142, 146, 152, 154, 156, 162, 178, 184$ *are in $K2 \cup K3$.*

Proof. We are going to prove the statement for ECA rule 2. For all other rules, the argument is quite similar. We prove that the fixed point $\begin{bmatrix} 0 \\ 0 \end{bmatrix}$ is not an equicontinuity point for the RESOCA 2. For that, we are going to build an arbitrary close point such that its orbit diverges from the fixed point. Fix some ϵ and let $\delta \leq \epsilon$ (if $\delta > \epsilon$ then we are done). Let n be the least integer such that $2^{-n} < \delta$. Consider a configuration $u \in \{0,1\}^{\mathbb{Z}}$ such that $u_i = 1$ if $i = n$ and $u_i = 0$ otherwise. Now build $v = \begin{bmatrix} 0 \\ u \end{bmatrix}$. It is not difficult to see that $G^n_{\delta'_2}(v)_{n-t} = \begin{bmatrix} 0 \\ 1 \end{bmatrix}$ for $t \in \mathbb{N}$. Hence, $d\left(\begin{bmatrix} 0 \\ 0 \end{bmatrix}, G^n_{\delta'_2}(v) \right) = 1$.

Fig. 6. The ECA rule 56 with initial configuration $\underline{1}$ (left) and its induced RESOCA with initial configuration $[\underline{1}, \underline{1}]^T$ (right). Time grows downward. Only the second component of each configuration is plotted.

With similar techniques as in Proposition 4 (but for different equicontinuity points) one can prove the following.

Fig. 7. The ECA rule 105 with initial configuration $\underline{1}$ (left) and its induced RESOCA with initial configuration $[\underline{1}, \underline{1}]^T$ (right). Time grows downward. Only the second component of each configuration is plotted.

Proposition 5. *The RESOCA induced by ECA rules* $5, 7, 9, 11, 13, 15, 25, 29,$ $33, 35, 37, 41, 43, 45, 57, 73, 77, 105$ *are in K2 $\cup$ K3.*

Figures 6 and 7 provide space-time diagrams for representative rules described in Propositions 4 and 5.

We say that a HOCA of local rule δ' is **periodic in the second component** if for any $c \in \{0,1\}^{\mathbb{Z}}$, the sequence $c, G_{\delta'}(c)_2, G_{\delta'}^2(c)_2, \ldots$ is periodic. The following lemma connects periodicity in the second component with periodicity of the global function. Its proof is trivial.

Lemma 8. *If a HOCA is periodic in the second component iff it is periodic.*

Proposition 6. *RESOCA 51 is periodic. Hence, it is in K1.*

Proof. We are going to prove that RESOCA 51 is periodic in the second component. Lemma 8 will allow to conclude. For any $c \in \{0,1\}^{\mathbb{Z}}$ and any $t \in \mathbb{N} \setminus \{0\}$, according to the local rule of ECA 51 we have that

$$(c_2^{t+1})_i = 1 + (c_2^t)_i + (c_2^{t-1})_i \quad \text{mod } 2.$$

Hence, at time $t + 2$ we have

$$(c_2^{t+2})_i = 1 + (c_2^{t+1})_i + (c_2^t)_i = 1 + 1 + (c_2^t)_i + (c_2^{t-1})_i + (c_2^t)_i = (c_2^{t-1})_i.$$

We conclude that $c, G_{\delta'_{51}}(c)_2, G_{\delta'_{51}}^2(c)_2, \ldots$ is periodic with period 2.

The following result needs a completely different proof technique than the previous one.

Proposition 7. *The RESOCA induced by ECA rule 1 periodic. Hence, it is in K1.*

5   Conclusions and Perspectives

The research work investigated the dynamics of a specific type of second-order elementary cellular automata up to topological conjugacy. The notion of transformation graph has been adapted to the new context and the index of the

Table 3. RESOCA rules that still need to be fully investigated.

3, 4, 8, 12, 19, 23, 27, 32, 36, 40, 44, 72, 76, 104, 108, 128, 132, 136, 140, 160, 164, 168, 172, 200, 232

equivalence relation induced by a specific set of local rule transformations has been computed. The remark that this index is too huge for being able to provide a complete characterisation in the present paper induced us to the introduction of a (considerably) restricted set of second order CA the we called RESOCA . The paper puts forth significant derivations related to RESOCA , in terms of equicontinuity, periodicity, expansivity, and strong transitivity. However, even this restricted class proved to be quite complicated to study and the dynamics of 25 (see Table 3) out of 88 RESOCA rules have not been fully explored yet. Future research is needed to explore more complex types of second-order cellular automata that have not yet been investigated.

Appendix

Proof. (*Proof of Proposition 5*). We are going to prove the statement for ECA rule 5. For all other rules, the argument is quite similar. It is not difficult to prove that if the initial condition is $(\underline{1}, \underline{1})$, then for any $i \in \mathbb{N}$ the second component of the current configuration c_2^{3i} at time $3i$ is such that $(c_2^{3i})_k = 1$ if $k \in \{-i, \ldots, 0, \ldots, i\}$ and $(c_2^{3i})_k = 0$ otherwise.

Proof. (*Proof of Lemma 6*). Consider the configurations $\underline{1}$ and $\underline{0}$. Let δ be the local rule of an injective CA. Then, we can have two cases:

$G_\delta(\underline{1}) = \underline{0}$ and $G_\delta(\underline{0}) = \underline{1}$: in this case, chose a configuration z different from $\underline{1}$ and $\underline{0}$. Let $\varepsilon = \min(d(\underline{1}, \underline{0}), d(\underline{1}, z), d(\underline{0}, z))$. Then, in $B_\varepsilon(z)$ no point z' and time $t \in \mathbb{N}$ can be chosen such that $G_\delta^t(z') = \underline{1}$ without contradicting the injectivity;

$G_\delta(\underline{1}) = \underline{1}$ and $G_\delta(\underline{0}) = \underline{0}$: in this case one just need to chose $z = \underline{0}$ and $\varepsilon < d(z, \underline{1})$ and make a similar reasoning as in the previous item.

So in neither case the CA can be strongly transitive. It is well-known that expansivity implies strong transitivity. Hence, an injective CA cannot be expansive since it cannot be strong transitive.

Let us prove that injective CAs must have DPO. Indeed, by shift-invariance, the image of spatial periodic configuration is a spatial periodic configuration. By injectivity, any spatial periodic configuration is periodic point. By recalling that the set of spatial periodic configurations is dense in the set of configurations we have the thesis.

Proof. (*Proof of Lemma 7*) This is an immediate consequence of the fact that injective CA are surjective and of [1, Corollary 3.1].

Proof. (Proof of Proposition 7). Consider the ECA rule 1, it is such that if the neighborhood contains $(0,0,0)$ then the output is 1, 0 otherwise. Hence, δ_1' is such that

$$G_{\delta_1'}\left(\begin{bmatrix} c_1^t \\ c_2^t \end{bmatrix}\right)_i = \begin{bmatrix} (c_2^t)_i \\ 1-(c_1^{t-1})_i \end{bmatrix}$$

if $((c_2^t)_{i-1}, (c_2^t)_i, (c_2^t)_{i+1}) = (0,0,0)$; otherwise we have

$$G_{\delta_1'}\left(\begin{bmatrix} c_1^t \\ c_2^t \end{bmatrix}\right)_i = \begin{bmatrix} (c_2^t)_i \\ (c_1^{t-1})_i \end{bmatrix}$$

It is not difficult to see that $\begin{bmatrix} a \\ b \end{bmatrix}$ with $a,b \in \{11,01,10\}$ is a blocking word for δ_1'. Therefore, if one wants to try to build a configuration not containing blocking word one should consider the blocks

$$\begin{bmatrix} 11 \\ 00 \end{bmatrix}, \begin{bmatrix} 00 \\ 11 \end{bmatrix}, \begin{bmatrix} 01 \\ 00 \end{bmatrix}, \begin{bmatrix} 00 \\ 01 \end{bmatrix}, \begin{bmatrix} 10 \\ 00 \end{bmatrix}, \begin{bmatrix} 00 \\ 10 \end{bmatrix}, \begin{bmatrix} 00 \\ 00 \end{bmatrix}$$

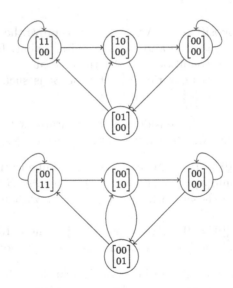

Fig. 8. Right extensions built from the blocks in the proof of Proposition 7.

Therefore a configuration which is potentially without blocking words can be described as the right extension obtained by the previous blocks and which can be represented by the union of the automata in Fig. 8. From the right extensions we immediately deduce that configurations without blocking words have $\underline{0}$ either as first or as second component. At this point let us prove that $w = \begin{bmatrix} 000 \\ 000 \end{bmatrix}$ is a

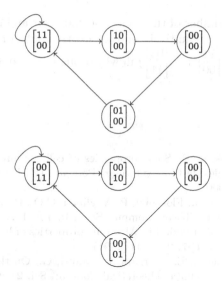

Fig. 9. Simplification of right extensions built from the blocks in the proof of Proposition 7.

blocking word. Indeed, assume that w is embedded in a configuration at some time $t > 0$ which has $\underline{0}$ as a second component. Then, we have

$$G_{\delta_1'}\left(\begin{bmatrix} x0.0.0y \\ 000.0.000 \end{bmatrix}\right) = \begin{bmatrix} 000.0.000 \\ \bar{x}1.1.1\bar{y} \end{bmatrix}$$

and

$$G_{\delta_1'}\left(\begin{bmatrix} 000.0.000 \\ \bar{x}1.1.1\bar{y} \end{bmatrix}\right) = \begin{bmatrix} \bar{x}1.1.1\bar{y} \\ 00.0.00 \end{bmatrix}$$

and finally

$$G_{\delta_1'}\left(\begin{bmatrix} \bar{x}1.1.1\bar{y} \\ 00.0.00 \end{bmatrix}\right) = \begin{bmatrix} 00.0.00 \\ 0.0.0 \end{bmatrix}$$

We conclude that the loop around $\begin{bmatrix} 00 \\ 00 \end{bmatrix}$ in the automata of Fig. 8 has to be deleted. In a much more painful and tedious way one can prove that also $\begin{bmatrix} 010 \\ 000 \end{bmatrix}$ and $\begin{bmatrix} 000 \\ 010 \end{bmatrix}$ (resp. $\begin{bmatrix} 101 \\ 000 \end{bmatrix}$ and $\begin{bmatrix} 000 \\ 101 \end{bmatrix}$) are blocking words. These facts imply that automata in Fig. 8 can be simplified as in Fig. 9. From 9 we deduce that target configurations might contain the block $\begin{bmatrix} 10.0.1 \\ 00.0.0 \end{bmatrix}$ (resp., $\begin{bmatrix} 00.0.0 \\ 10.0.1 \end{bmatrix}$) but (with some extra pain) one can prove that also this is a blocking word. This fact tels that in the right extensions that we were building the only configurations that can be built are $[\underline{0}, \underline{1}]$ or $[\underline{1}, \underline{0}]$ but both of the are periodic of period 2. We conclude by remarking that all blocking words are periodic and that any configuration contains an infinite number of them both on the right and on the left of the cell of index 0 unless they definitively end up on the right or on the left with an infinite repetition of $\begin{bmatrix} 11 \\ 00 \end{bmatrix}$ (or $\begin{bmatrix} 00 \\ 11 \end{bmatrix}$) in which but those configurations are easily seen to be periodic.

References

1. Blanchard, F., Tisseur, P.: Some properties of cellular automata with equicontinuity points. Annales de l'Institut Henri Poincare (B) Probability and Statistics **36**(5), 569–582 (2000)
2. Braga, G., Cattaneo, G., Flocchini, P., Vogliotti, C.Q.: Pattern growth in elementary cellular automata. Theor. Comput. Sci. **145**(12), 1–26 (1995)
3. Le Bruyn, L., Van den Bergh, M.: Algebraic properties of linear cellular automata. Linear Algebra Appl. **157**, 217–234 (1991)
4. Cattaneo, G., Formenti, E., Margara, L., Mauri, G.: On the dynamical behavior of chaotic cellular automata. Theoretical Comput. Sci. **217**(1), 31–51 (1999)
5. Cattaneo, G., Formenti, E., Margara, L., Mauri, G.: Transformations of the one-dimensional cellular automata rule space. Parallel Comput. **23**(11), 1593–1611 (1997)
6. Karel Culik, I.I., Sheng, Yu.: Undecidability of CA classification schemes. Complex Syst. **2**(2), 177–190 (1988)
7. Dennunzio, A., Formenti, E., Grinberg, D., Margara, L.: Chaos and ergodicity are decidable for linear cellular automata over $(\mathbb{Z}/m\mathbb{Z})^n$. Inf. Sci. **539**, 136–144 (2020)
8. Dennunzio, A., Formenti, E., Manzoni, L., Margara, L., Porreca, A.E.: On the dynamical behaviour of linear higher-order cellular automata and its decidability. Inf. Sci. **486**, 73–87 (2019)
9. Hedlund, G.A.: Endomorphisms and automorphisms of the shift dynamical system. Math. Syst. Theory **3**(4), 320–375 (1969)
10. Kari, J.: Linear cellular automata with multiple state variables. In: Reichel, H., Tison, S. (eds.) STACS 2000. LNCS, vol. 1770, pp. 110–121. Springer, Heidelberg (2000). https://doi.org/10.1007/3-540-46541-3_9
11. Kari, J.: Algorithms for group cellular automata. In: Das, S., Martínez, G.J. (eds.) ASCAT 2022. AISC, vol. 1425, pp. 17–25. Springer, Cham (2022). https://doi.org/10.1007/978-981-19-0542-1_2
12. Kůrka, P.: Languages, equicontinuity and attractors in cellular automata. Ergodic Theory Dynam. Systems **17**(2), 417–433 (1997)
13. Margolus, N.: Physics-like models of computation. Physica D **10**(1–2), 81–95 (1984)
14. Schüle, M., Stoop, R.: A full computation-relevant topological dynamics classification of elementary cellular automata. Chaos Interdiscipl. J. Nonlinear Sci. **22**(4), 043143 (2012)
15. Toffoli, T.: Computation and construction universality. J. Comput. Syst. Sci. **15**, 213–231 (1977)
16. Gérard YouYou Vichniac: Simulating physics with cellular automata. Physica D **10**(1–2), 96–116 (1984)

Author Index

M. Dalui et al. (Eds.): ASCAT 2024, CCIS 2021, p. 219, 2024.
https://doi.org/10.1007/978-3-031-56943-2